Sara Olibet

Interface properties of amorphous/crystalline silicon heterojunctions

Sara Olibet

Interface properties of amorphous/crystalline silicon heterojunctions

Modeling, experiments and solar cells

Südwestdeutscher Verlag für
Hochschulschriften

Imprint

Any brand names and product names mentioned in this book are subject to trademark, brand or patent protection and are trademarks or registered trademarks of their respective holders. The use of brand names, product names, common names, trade names, product descriptions etc. even without a particular marking in this work is in no way to be construed to mean that such names may be regarded as unrestricted in respect of trademark and brand protection legislation and could thus be used by anyone.

Cover image: www.ingimage.com

Publisher:
Südwestdeutscher Verlag für Hochschulschriften
is a trademark of
Dodo Books Indian Ocean Ltd. and OmniScriptum S.R.L publishing group

120 High Road, East Finchley, London, N2 9ED, United Kingdom
Str. Armeneasca 28/1, office 1, Chisinau MD-2012, Republic of Moldova, Europe
Managing Directors: Ieva Konstantinova, Victoria Ursu
info@omniscriptum.com

Printed at: see last page
ISBN: 978-3-8381-0971-8

Zugl. / Approved by: Neuchâtel (CH), University, PhD, 2008

Copyright © Sara Olibet
Copyright © 2009 Dodo Books Indian Ocean Ltd. and OmniScriptum S.R.L publishing group

Abstract

The main focus of this work is the study of interfaces in amorphous/crystalline silicon (a-Si:H/c-Si) heterostructures, especially the investigation of the a-Si:H/c-Si heterointerface's electronic quality and its effect on the consecutive a-Si:H/c-Si heterojunction (HJ) solar cell fabrication.

c-Si based solar cells have the potential for achieving high conversion efficiencies, but the standard simple fabrication processes lead to medium module efficiencies. Thin-film Si based technologies offer the prospect of low-cost fabrication but yield lower efficiencies. a-Si:H/c-Si HJ solar cells combine the advantages of both technologies, i.e., the high efficiency potential of c-Si and the low fabrication cost of a-Si:H. In this way the c-Si cost becomes reasonable because it is possible to use very thin wafers to produce highly efficient solar cells.

The electronic quality of the heterostructure interface was evaluated experimentally with photogenerated carrier lifetime measurements. In this study, carrier recombination at the interface is the step limiting photogenerated carriers' lifetime. In the theoretical part of this work, c-Si surface recombination is modeled by considering for the first time the amphoteric nature of Si dangling bonds. For this, a model previously established for bulk a-Si:H recombination, is extended to the description of the c-Si surface recombination through amphoteric defects. Its differences and similarities compared to existing interface recombination models are discussed. This new model is currently the simplest that allows an understanding of the largest set of experimentally observed behaviors of passivation layers on c-Si. The potential of the model's applicability to passivation by silicon dioxide (SiO_2) and silicon nitride (SiN_x) layers is also demonstrated.

The passivation performances of a-Si:H on c-Si are examined by growing symmetrical layers and layer stacks (intrinsic, microdoped, intrinsic plus doped) by very high frequency plasma enhanced chemical vapor deposition (VHF-PECVD, at 70 MHz). Lifetime measurements in combination with numerical modeling, incorporating our new amphoteric interface recombination model, reveal the microscopic passivation mechanism of a-Si:H on c-Si. The growth of intrinsic (i) a-Si:H efficiently reduces the c-Si dangling bond density at the crystallographic interface. It further decreases the interface recombination rate when set (by the wafer's doping level and type or by an outer potential) in a neutral state (with the smallest free carrier capture cross-sections). Furthermore, the magnitude of an additional field-effect passivation can be tuned by fixing the i a-Si:H's outer surface potential when capping it with a doped thin-film Si layer. i a-Si:H passivation implies complete devices with very high calculated open-circuit voltages (V_{OC}) over 700 mV on flat c-Si of all kinds of doping types and levels. This corresponds to effective record low surface recombination velocities under 5 cm/s (and down to 1 cm/s).

The emerging interpretation of lifetime measurements on specific heterostructure test samples allows for a rapid Si HJ solar cell development using a fast device diagnostic procedure, based on a single process step analysis. Individual testing of emitter and back surface field (BSF) layers on c-Si wafers allows for a rapid test of their suitability for Si HJ solar cell fabrication.

In order to verify the demonstrated passivation quality and suitability of emitter and BSF layer stacks, Si HJ solar cells are produced. On flat n-type c-Si, good results are rapidly achieved, i.e. V_{OC}s up to 715 mV and efficiencies up to 19.1%. On flat p-type c-Si, V_{OC}s up to 690 mV and efficiencies up to 16.3% are reached. On textured c-Si, the a-Si:H's passivation capability depends on the wafers' surface morphology. Transmission electron microscopy (TEM) micrographs of the textured thin-film Si/c-Si interface, shown here for the first time, shows local epitaxial growth of i a-Si:H in c-Si valleys. Lifetime measurements make it possible to attribute the cause of an increased interface recombination to these features. High-quality texture achieves the same high implied V_{OC}s by i a-Si:H passivation as flat c-Si, but interface recombination is still increased on standard textured c-Si. Decreasing the density of epitaxized i-layers by using a large pyramidal texture, a modified BSF layer growth and an additional surface morphology modification, yields complete textured cells with very high V_{OC} values over 700 mV.

Contents

1 Introduction   **1**
1.1 General context and objective 1
1.2 Structure of the document 2
1.3 Contribution of this work to the research field 3

2 Experimental   **5**
2.1 Measurement techniques 5
    2.1.1 Layer characterization 6
        2.1.1.1 Thickness 6
        2.1.1.2 Optical transmission, reflectivity and absorption 6
        2.1.1.3 Conductivity 8
        2.1.1.4 Crystallinity 8
    2.1.2 a-Si:H/c-Si heterostructure characterization 10
        2.1.2.1 a-Si:H/c-Si interface recombination quantification: lifetime measurements 12
        2.1.2.2 Nanometrically resolved imaging of surfaces, interfaces and layer structures 16
        2.1.2.3 Subnanometrically resolved imaging of interfaces and layer structures 17
        2.1.2.4 Solar cell efficiency measurement: current-voltage characteristic 19
        2.1.2.5 External quantum efficiency 21
        2.1.2.6 Acquiring series-resistance-less JV-curves: SunsV_{OC} measurements 23
2.2 Pre-deposition wafer treatment 24
    2.2.1 Native oxide removal 25
    2.2.2 Textured c-Si cleaning issue 27
2.3 VHF-PECVD deposition of amorphous and microcrystalline silicon 28
    2.3.1 VHF-PECVD deposition chamber 29

		2.3.2	Intrinsic amorphous silicon	31

        2.3.2   Intrinsic amorphous silicon 31
        2.3.3   (Intrinsic) amorphous silicon microdoping 31
        2.3.4   Doped microcrystalline silicon 32
  2.4   Contact formation 34
        2.4.1   Transparent conducting contact deposition: ITO . . 35
        2.4.2   Metallization 40
  2.5   Full a-Si:H/c-Si heterostructure processing 41
        2.5.1   a-Si:H/c-Si passivation samples 41
        2.5.2   a-Si:H/c-Si heterojunction solar cells 42

3 Heterostructure interface recombination modeling   **45**
  3.1   Introduction 45
  3.2   Bulk recombination modeling: τ_{Aug}, τ_{rad} and τ_{defect} 47
  3.3   Standard interface recombination modeling 50
        3.3.1   Shockley-Read-Hall interface recombination 50
        3.3.2   Extended Shockley-Read-Hall interface recombination formalism 56
        3.3.3   Emitter and BSF recombination: double-diode modeling 62
  3.4   Determination of interface recombination parameters: interface recombination center density, field-effect passivation . 64
        3.4.1   Issues when comparing modeled and measured injection level dependent recombination curves 64
        3.4.2   Modeling the standard c-Si surface passivation schemes SiO_2 and SiN_x 68
                3.4.2.1   Example SiO_2 68
                3.4.2.2   Example SiN_x 71
  3.5   Novel model for a-Si:H/c-Si interface recombination based on the amphoteric nature of silicon dangling bonds 73
        3.5.1   Introduction 73
        3.5.2   a-Si:H bulk recombination 74
        3.5.3   Extension to a-Si:H/c-Si interface recombination . . 85
  3.6   Conclusion: comparison of the different interface recombination schemes 93

4 a-Si:H/c-Si interface passivation: experiment & modeling 101
  4.1   Experiment and modeling 101
  4.2   State of the art Si surface passivation 104
  4.3   Intrinsic a-Si:H on various flat c-Si substrates 106
        4.3.1   Hardware and physical effects affecting the measurements 112

	4.4	Intrinsic a-Si:H of varying thicknesses	116
		4.4.1 a-Si:H thickness dependent passivation	116
		4.4.2 Light degradation	118
		4.4.3 Dark degradation	121
	4.5	Additional field-effect passivation	125
		4.5.1 Microdoped a-Si:H	125
		4.5.2 Stacks of intrinsic a-Si:H plus doped a-Si:H/µc-Si:H	130
	4.6	Atomic structure of the a-Si:H/c-Si heterointerface	136
	4.7	Influence of the texture morphology	141
		4.7.1 Intrinsic a-Si:H passivation	145
		4.7.2 Emitter and BSF layer stack passivation	151
	4.8	Limits imposed on V_{OC} and FF by interface recombination: choice of the optimal c-Si doping type and level for Si HJ solar cell fabrication	157
	4.9	Conclusions on optimized a-Si:H/c-Si interface passivation	162
5	**Amorphous/crystalline silicon heterojunction solar cells**		**165**
	5.1	Introduction	165
	5.2	Carrier transport in a-Si:H/c-Si heterojunction solar cells .	167
	5.3	Lifetime measurements as a guide for solar cell optimization	171
	5.4	a-Si:H/c-Si heterojunction solar cells based on flat c-Si . .	175
	5.5	Textured a-Si:H/c-Si heterojunction solar cells	179
		5.5.1 Introduction	179
		5.5.2 Influence of the texture morphology	184
		5.5.3 Pyramidal valley rounding	188
	5.6	Conclusions on amorphous/crystalline silicon heterojunction solar cells	196
6	**Summary, conclusions and further work**		**199**
Acknowledgments			**203**
Glossary			**204**
Bibliography			**212**
List of publications			**233**
A	**Numerical surface potential calculation**		**235**
	A.1	Surface potential calculation: numerical approximation of Ψ_s from Q_{it}, Q_f and Q_G	235

Contents

A.2 Surface potential calculation: numerical solution of non-linear equation relating Ψ_s and Q_s 237

Chapter 1

Introduction

1.1 General context and objective

The photovoltaic effect, i.e. the conversion of solar light energy into electricity, was discovered by the French scientist A. E. Becquerel in 1839 [Bec39]. However, it was not until 1954, that Chapin *et al.* achieved a sunlight energy conversion efficiency of 6 percent in the Bell laboratories [CFP54]. The solar cell devices first gained interest as an energy supply in space applications. As for all renewable energies, the current driving force for the boom in the photovoltaic (PV) market is the exhaustion of fossil fuel energy. Total production of PV was 3.8 GW worldwide in 2007, having grown by an average of 48% each year since 2002 [ear]. The top five PV-producing countries are China, Japan, Germany, Taiwan and the United States. The top ten PV-producing companies are Q-Cells, Sharp, Suntech, Kyocera, First Solar, Motech, SolarWorld, Sanyo, Yingli and JA Solar [wor].

The PV market is dominated by silicon (Si) wafer-based solar cells, having 90% of market share. Although silicon is abundant, there is a crystalline silicon (c-Si) shortage, ever since PV needs have surpassed the left-over from computer chip fabrication in 2005. However, considering the energy required for the standard crystalline silicon solar cell's fabrication, the industrial module conversion efficiency is still too low at typically 12 to 14%. Contrariwise, thin-film based technologies such as principally thin silicon layers, i.e. amorphous and microcrystalline silicon (a-Si:H and µc-Si:H), offer the prospect of processing large areas at extremely low-cost. But so far only moderate efficiencies of 6 to 8.5% have been achieved by large-scale fabricated modules. The best of both technologies is combined in the amorphous/crystalline silicon heterojunction (a-Si:H/c-Si HJ) solar cells: the high efficiency of c-Si based solar cells and the low-cost solar cell processing of thin-film Si. Because of low process temperatures and the absence of a thick aluminum layer on the back, very thin Si wafers can be used and thus the c-Si cost becomes reasonable, while maintaining high efficiencies thanks to a very efficient suppression of interface recombination losses at the a-Si:H/c-Si interface. In 1990 the company Sanyo [san] started its research in the field of Si heterojunctions and launched mass production of their so-called "HIT" (heterojunction with intrinsic thin-layer) cells in 1997. Sanyo's HIT cells efficiencies are outstanding: 19% on the cell and 17% on the module level in production, together with a world record laboratory solar cell conversion efficiency of 22.3% on 100.5 cm^2 [TYT$^+$09]. Sanyo's market share is actually 5% and it aims to reach 1 GW production in 2010, followed by 4 GW in 2020, targeting a total PV market share of 10%.

The key feature for the high efficiency of Si HJ solar cells is the excellent surface passivation of amorphous Si on c-Si. Given the excellent quality of today's monocrystalline Si wafers, charge carrier recombination losses occur principally at the c-Si's surface. The strongly reduced charge carrier recombination at the a-Si:H/c-Si heterointerface yields outstanding open-circuit voltages (V_{OC}) of 740 mV achieved even on 80 μm thick Si HJ solar cells [TYT$^+$09]. Such V_{OC}s are virtually out of reach for standard c-Si solar cell processing, as the passivation layer deposited on top of the diffused emitter has to be locally opened (i.e. depassivated) to draw current from the solar cell.

Despite these excellent achievements, the physical understanding of interfaces in amorphous/crystalline silicon heterojunctions is currently limited. Therefore, while the passivation mechanism of the standard c-Si surface passivation schemes silicon dioxide (SiO_2) and silicon nitride (SiN_x) are well known, the surface passivation mechanism of amorphous and microcrystalline silicon layers and layer stacks on c-Si is relatively new and will be studied in detail in the first part of this work. Only a few groups have attempted to reach some of the results attained by Sanyo, but the technological understanding is also strongly limited. Thus, the aim of the second part of this work is to contribute to the device development.

1.2 Structure of the document

First, the measurement and deposition techniques used in the study of the interface properties of amorphous/crystalline Si heterojunctions are presented together with the Si heterostructure fabrication processes (chapter 2). The advanced reader may skip this part and come back later if needed.

Surface recombination reduction relies on a combination of surface defect decrease and field-effect passivation, differently weighted for different surface passivation schemes. To determine the individual contribution of these two effects for the a-Si:H/c-Si interface passivation scheme, we establish a new model for a-Si:H/c-Si interface recombination based on the amphoteric nature of silicon dangling bonds (chapter 3). Numerical modeling is compared to various experimentally measured injection level dependent charge carrier lifetimes. Various combinations of intrinsic (i), microdoped or internally polarized i a-Si:H layers are used on a wide set of wafers with varying doping levels and types (p, n and intrinsic) (chapter 4). Interface recombination is directly related to the solar cell's V_{OC} and also sets upper limits on the fill factor that a solar cell with a specific interface passivation scheme can reach. Lifetime studies on intrinsic plus doped thin-film Si layer stacks serve as a prerequisite for highly efficient Si HJ solar cell fabrication (chapter 4&5). Chapter 4 as well as chapter 5 are supposed to be self-consistent. When growing intrinsic / doped thin-film Si layer stacks on c-Si to form HJ solar cells, the c-Si's surface passivation takes place right at the junction. While this has the advantage of avoiding a direct contact between the metal and the electrically active semiconductor, the requirements on the c-Si's surface quality become higher as will become obvious from the results obtained for our textured Si HJ solar cells. In this case, transmission electron microscopy (TEM) micrographs permit the identification of the microstructured features causing limited device performances (chapter 5).

1.3 Contribution of this work to the research field

In the last years the research on Si heterojunctions has gained in importance, and work performed by other authors is considered in the following chapters. This PhD thesis contributes to the research field with the following elements:

- For the first time the amphoteric nature of Si dangling bonds is taken directly into account for the modeling of a-Si:H/c-Si interface recombination ([OliPRB07], Sec. 3.5).

- An original representation in terms of trajectories over surface recombination rate surface plots facilitates the intuitive interpretation of

1.3. Contribution of this work to the research field

injection level dependent charge carrier lifetime curves ([OliNREL07], Chap. 3).

- The amphoteric Si interface recombination model with its characteristic local minimum also permits interface recombination measurements made at the SiO_2/c-Si interface to be fitted (Sec. 3.6).

- Our new amphoteric interface recombination model allows a satisfactory quantitative description of the measured injection level- dependance of interface recombination, revealing the microscopic passivation mechanism of a-Si:H on c-Si ([OliPRB07], Chap. 4).

- In view of maximal Si HJ solar cell performances with a maximal tolerance to an increased interface defect density, the identified a-Si:H/c-Si interface passivation mechanism implies the choice of the optimal wafer type (Sec. 4.8).

- The emerging interpretation of lifetime measurements on specific heterostructure test samples allows for a rapid Si HJ solar cell development based on a single process step analysis ([OliNum07], Sec. 5.3).

- a-Si:H/c-Si HJ solar cell devices with high open-circuit voltages and high efficiencies are demonstrated (small surface) ([OliDresd06,FesMil07, OliFuk07], Chap. 5).

- Efficiency limiting factors such as e.g. inappropriate texture morphologies, epitaxial growth of the passivating layer or an insufficient back surface field are identified ([OliVal08], Sec. 4.7, Sec. 4.6, Sec. 5.5).

Chapter 2

Experimental

In this chapter the measurement and deposition techniques used in this study of the interface properties of amorphous/crystalline Si heterojunctions are presented. Information on the detailed process flow ranging from taking a c-Si wafer out of its box to a finished Si heterojunction solar cell is described.

2.1 Measurement techniques

Thin-film Si layers, layer stacks and contact layers for further use in a-Si:H/c-Si heterostructures are first individually developed on glass. On one hand, this offers a simpler processing and handling. On the other hand, not all electronic and optical measurements can be done on c-Si because of the substrate influence masking the signals of the thin layers' properties. The interface recombination properties of the previously developed thin-film Si layers and layer stacks can be excellently characterized by means of effective carrier lifetime measurements. These directly probe the heterointerface's recombination when high-quality c-Si wafers are used for the films deposition. The microstructure and atomic nature of interfaces in a-Si:H/c-Si heterostructures are visualized by means of scanning and transmission electron microscopy (SEM and TEM). While layer development on glass is very useful for flat Si heterojunctions, these optimizations cannot necessarily be transferred to textured wafers. Finally, Si heterojunction solar cell performances are evaluated by means of current-voltage characteristic (JV-curve), quantum efficiency (QE) and illumination level dependent V_{OC} (SunsV_{OC}) measurements.

2.1.1 Layer characterization

Single and stacked thin-film silicon and contact layers are grown on glass (Schott AF45) for thickness, electronic and optical characterization. While thickness measurements are performed on glass for convenience, coplanar electronic and optical characterization needs to be done on glass to separate c-Si bulk properties from the properties of the overlying thin-film layers.

2.1.1.1 Thickness

The layer thickness is measured using an alpha-step profiler (Ambios XP-2) allowing layer thickness measurements below 15 nm. A suitable step for profiling is obtained as follows:

- For thickness measurements of thin silicon layers on glass a step is formed by exploiting the selective etching of silicon over glass, which may be accomplished by one of the two following methods:
  - KOH etching of holes into the thin silicon layer,
  - masking the thin silicon film with a marker and then etching away the surrounding silicon by an SF_6/O_2 plasma attack. Afterwards, the marker is dissolved in acetone.

- For the contact layers' thickness measurement (typically around 100 nm), the substrate is marked with a marker before the layer deposition. After deposition, these marks (including the deposited layer on them) are dissolved in acetone leaving back bare glass stripes.

The growth rate of a layer is then determined by dividing this measured film thickness by the deposition time.

The alpha-step profilers lower detection limit is 10 nm with an error of ± 3 nm. Even thinner layer thicknesses can be accurately measured by ellipsometry [FW07]. To measure the thickness of layers deposited on textured substrates, the only means is by examination of cross-sections with SEM (Sec. 2.1.2.2) or for ultra-thin layers, by TEM micrographs (Sec. 2.1.2.3).

2.1.1.2 Optical transmission, reflectivity and absorption

The total wavelength (λ [nm]) dependent transmission ($Tr(\lambda)$ [%]) and reflection ($Refl(\lambda)$ [%]) are acquired using a double-beam Perkin Elmer Lambda 900 UV/Vis/NIR photospectrometer with an integrating sphere.

2.1. Measurement techniques

Layers deposited on glass are measured with the layer surface towards the incoming light beam. From this the absorption spectrum ($A(\lambda)$ [%]) can be calculated, because all light that is neither transmitted nor reflected is absorbed:

$$A(\lambda) = 100 - Tr(\lambda) - Refl(\lambda). \tag{2.1}$$

Fig. 2.1 shows such an optical transmission and reflection measurement and the corresponding absorption curve for a 85 nm thick indium tin oxide (ITO) layer on glass (Fig. 2.1(a)) and the amorphous/microcrystalline (a-Si:H/µc-Si:H) emitter and back surface field (BSF) layer stacks on glass (Fig. 2.1(b)).

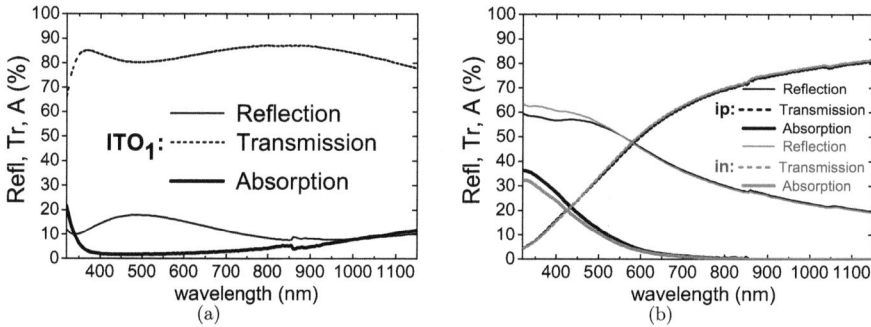

Figure 2.1: *Optical transmission, reflection and absorption of a) 85 nm ITO and b) a-Si:H/µc-Si:H emitter and BSF layer stacks on glass.*

Transmission and reflection measurements also allow the determination of the absorption coefficient α [cm^{-1}] as a function of the wavelength for large values of α. Sub-bandgap absorption has to be determined by other techniques such as by photothermal deflection spectroscopy (PDS) [WAC80] or by the constant photocurrent method (CPM) [VKST81]. Figure 2.2 summarizes the wavelength dependent absorption coefficients of amorphous, microcrystalline and monocrystalline silicon.

The absorption in a material is linked to the incident light intensity penetrating the sample I_L [mWcm^{-2}] by:

$$I(x) = I_L e^{-\alpha x}, \tag{2.2}$$

where x is the distance from the incident surface. The light penetration depth is therefore the inverse of the absorption coefficient α, and is plotted on the righthand ordinate of Fig. 2.2.

7

2.1. Measurement techniques

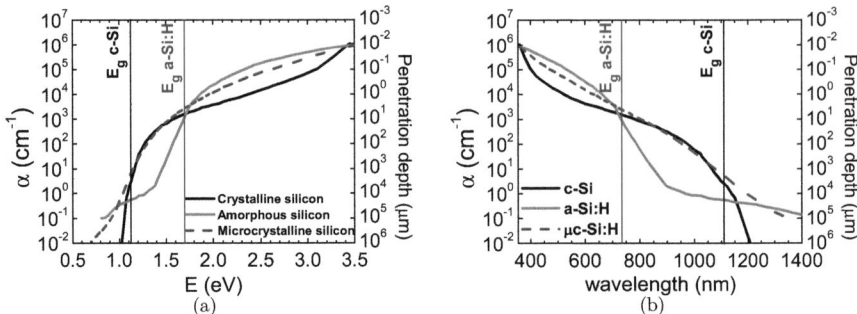

Figure 2.2: *Absorption coefficient of amorphous, microcrystalline and crystalline silicon as a function of a) the photon energy and b) the wavelength [SSV+04]. The righthand ordinate shows the light penetration depth in the different Si materials. The bandgap of c-Si and a-Si:H is indicated.*

2.1.1.3 Conductivity

The coplanar conductivity of doped μc-Si:H and contact layers is measured by a four point probe setup. The four probes arranged on a line measure the resistivity of samples by passing a current through the two outer probes and measuring the voltage drop through the inner probes. However, doped a-Si:H layers are too resistive to be measured by this setup. For this, coplanar ohmic contacts consisting of a 100 nm thick aluminum layer are thermally evaporated onto the doped a-Si:H layer. For the latter type of sample, the dark-conductivity activation energy E_{act} [eV] can also be measured. It is related to the energy difference from the Fermi level to the current-transporting energy band. The dark-conductivity σ_d [$(\Omega cm)^{-1}$] follows an exponential increase with temperature (T [K]):

$$\sigma_d(T) = \sigma_0 e^{\frac{-E_{act}}{kT}}. \tag{2.3}$$

Samples are heated up to 180 °C at a pressure of 10 mbar in an inert nitrogen atmosphere, maintained at 180 °C during 1 hour and then slowly cooled down. The plot of $\ln\sigma_d(T)$ vs $\frac{1}{T}$ is then used to extract E_{act}.

2.1.1.4 Crystallinity

Microcrystalline silicon is composed of conglomerates of nanocrystals embedded in amorphous tissue [VSKM+00]. A representative value of the

2.1. Measurement techniques

crystalline volume fraction of a silicon thin-film layer can be obtained by micro-Raman spectroscopy. Using this method, monochromatic light is focused on the layer through a conventional microscope that also collects the backscattered light. The inelastic scattering results from the interaction of light and matter as first predicted by Raman et al. [RK28]. Most of the scattered photons have the same energy (and thus wavelength) as the incident photons. However, a very small fraction ($\sim 1 \times 10^{-6}$) of the scattered light interacts inelastically with vibrational modes and thus looses or gains energy. The frequency shift observed in Raman scattering with the Renishaw Raman Imaging Microscope, system 2000, is characteristic of the chemical bonds present in the material. Figure 2.3 shows the Raman spectrum of a layer near the a-Si:H/µc-Si:H transition.

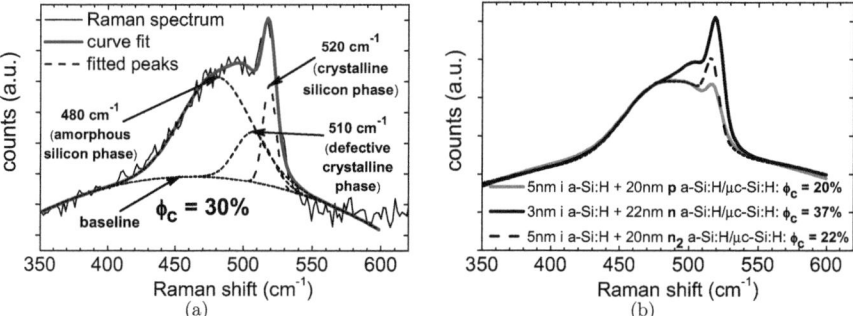

Figure 2.3: *a) Evaluation of the Raman crystalline fraction of a Si thin-film layer near the a-Si:H/µc-Si:H transition (20 nm a-Si:H/µc-Si:H on 5 nm intrinsic (i) a-Si:H on glass) by deconvolution of the Raman spectrum into three Gaussian peaks, as described by Eq. 2.4. b) Raman spectra of the emitter layer i/p stack and the BSF layer i/n stacks used for Si heterojunction solar cell formation (Sec. 2.3.4).*

This Raman spectrum can be deconvoluted into three Gaussian peaks, in which the integrated peak areas are proportional to the corresponding phase concentrations:

- a broad peak centered at 480 cm^{-1}, characteristic of the transverse optical (TO) mode in a-Si:H,

- a peak centered at around 510 cm^{-1}, attributed to a contribution of the crystalline volume fraction (defective part of the crystalline phase, such as a nanocrystal surface),

- and a narrow peak centered at 520 cm^{-1}, corresponding to the TO mode in c-Si.

The Raman crystallinity factor ϕ_c [%] is defined here as the ratio of the integrated area of the peaks related to the crystalline parts over the total area of silicon-related peaks:

$$\phi_c = \frac{I_{520} + I_{510}}{I_{520} + I_{510} + I_{480}}. \tag{2.4}$$

This factor does, however, not reflect the actual crystalline volume fraction that must take in consideration the integrated Raman cross-sections [VSDM+06].

When using green laser excitation light (514 nm wavelength) the crystallinity of layers with thicknesses as thin as 15 nm can still be evaluated due to the green light's high absorption coefficient in a-Si:H and µc-Si:H, as described in Fig. 2.2(b).

2.1.2 a-Si:H/c-Si heterostructure characterization

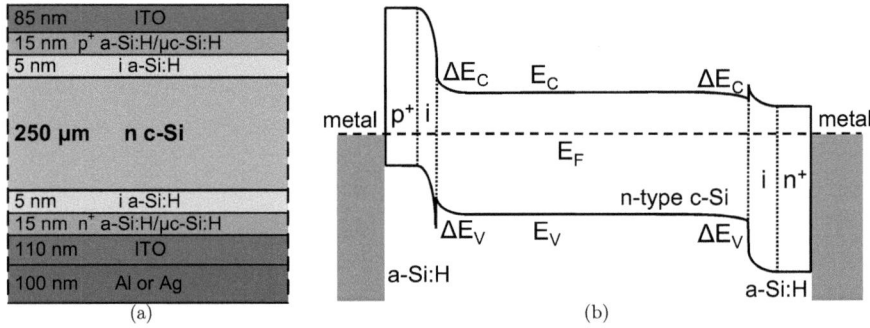

Figure 2.4: *a) Sketch and b) usual band diagram of an a-Si:H/c-Si heterojunction solar cell.*

Principally, an a-Si:H/c-Si heterostructure (Fig. 2.4 shows as an example the complete Si HJ solar cell) is dominated by both the c-Si bulk and its surface characteristics. For a good device, it is essential that the c-Si bulk features a low bulk recombination rate R [cm^{-3}s^{-1}]. Its importance in a device is best illustrated by defining the charge carrier lifetime $\tau \equiv \frac{\Delta n}{R}$ [s],

2.1. Measurement techniques

i.e., the average time it takes an excess carrier to recombine at an excess carrier density $\Delta n = \Delta p$ [cm^{-3}]. Photogenerated minority carriers need to diffuse to the junction to be collected so that they can contribute to the current. Therefore, the lifetime that determines the diffusion length L_{diff} [cm] by $L_{diff} = \sqrt{D\tau}$, where D [cm^{2}s^{-1}] is the diffusivity, is of utmost importance, as $L_{diff} > W$ [cm], where W is the wafer thickness, is the base condition for a high performance c-Si solar cell device.

When using high-lifetime float zone (FZ) wafers within a heterostructure, the device is dominated by its interface properties. The most important heterostructure interface feature is again a low interface recombination rate U [cm^{-2}s^{-1}]. Experimentally, the overall recombination rate can be evaluated by effective carrier lifetime measurements. The heterostructure's interface recombination properties can be directly accessed because in first approximation bulk recombination is negligible in high-lifetime wafers. For low heterostructure interface recombination, it has been shown that an atomically abrupt heterostructure interface is essential. High resolution (HR)-TEM micrographs are the best ways to visualize the heterostructure's interface qualities on the subnanometrical scale in operating devices.

The open-circuit voltage V_{OC} [V] of a finished solar cell is determined by its c-Si's bulk and surface recombination losses. To extract power from a functional solar cell device, photogenerated electrons and holes have to be successfully separated. While carrier extraction in short-circuit current (J_{SC} [mAcm^{-2}]) conditions is straightforward, series resistances, blocking junctions, shunts and diode non-idealities can hinder carrier extraction in the maximum power delivery operation conditions of a solar cell. The acquisition of illuminated current-voltage characteristics (JV-curves) can elucidate such problems and the solar cell efficiency can be calculated.

In fact, not all incident light reaches the c-Si absorber or can be absorbed therein before leaving the solar cell again. While the absorption coefficient of c-Si is well known, the complete solar cell's wavelength dependent losses due to reflection, transmission and absorption in non photoelectrically active layers, can be measured using external quantum efficiency (EQE [%]).

Finally, while JV-curves allow the calculation of a solar cell's efficiency under standard test conditions, many factors influence these characteristics, masking physical phenomena. The measurement of the V_{OC}'s illumination level-dependance (SunsV_{OC}) permits the evaluation of the upper cell efficiency limit imposed by recombination losses without the effect of series resistances as introduced by contacts and interfaces.

2.1. Measurement techniques

2.1.2.1 a-Si:H/c-Si interface recombination quantification: lifetime measurements

Carriers photogenerated in c-Si recombine in the c-Si bulk and at its surfaces, which form interfaces with a-Si:H. The time these carriers exist in the c-Si bulk before recombination is referred to as effective lifetime τ_{eff} [s] and is given by the overall lifetime in the c-Si bulk τ_{bulk} [s] and at its surface τ_{surf} [s]:

$$\frac{1}{\tau_{eff}} = \frac{1}{\tau_{bulk}} + \frac{1}{\tau_{surf}}. \tag{2.5}$$

Recombination in c-Si can be neglected in a first approximation when using high-quality float zone (FZ) grown wafers, i.e. $\tau_{bulk} \gg \tau_{surf}$. The experimentally accessible property is the (excess) conductivity induced in c-Si by the photogenerated excess carriers. This excess conductance can be sensed contactless by different techniques such as inductively by a coil [KS85], by microwave reflectance [DN62] or by the transmission of infrared light [BBB+02]. The quasi-steady-state photoconductance (QSSPC) and the microwave-detected photoconductance decay (MW-PCD) technique have found widespread use. More recently infrared camera lifetime mapping (ILM) has gained interest because within a short time space-resolved lifetime measurements can be made [BKBS00,PB04]: periodically photogenerated excess carriers in a heated wafer modulate the emitted infrared radiation that is observed with a camera. Also photoluminescence (PL) measurements permit lifetime measurements, as PL is directly related to the separation of the quasi-Fermi levels that in their turn are given by the excess carrier density [TBH+04]. In this study, the QSSPC technique was extensively used and some samples were also characterized by ILM at the Institut für Solarenergieforschung Hameln/Emmerthal (ISFH).

PCD and QSSPC measurements: the Sinton lifetime tester The electronic properties of silicon wafers and their surfaces are investigated using the photoconductance decay (PCD) and the quasi-steady-state photoconductance (QSSPC) techniques [SC96] with the WCT-100 photoconductance tool from Sinton Consulting [sin]. To ensure a homogeneous carrier generation throughout the whole c-Si bulk, a filter mounted on the flash lamp provides infrared illumination. In this setup, the inductively measured excess photoconductance $\Delta\sigma$ [$(\Omega cm)^{-1}$] is given by

$$\Delta\sigma = q(\Delta n_{av}\mu_n + \Delta p_{av}\mu_p)W, \tag{2.6}$$

where $\Delta n_{av} = \Delta p_{av}$ is the average excess carrier density (Δn) [cm^{-3}], W [cm] the wafer thickness and μ_n, μ_p [cm^{2}V^{-1}s^{-1}] the electron and hole

2.1. Measurement techniques

mobilities in c-Si whose values are well known.

The transient PCD measurement mode consists of measuring wafer conductivity vs time after a very short and intense light pulse. The effective carrier lifetime τ_{eff} at each excess carrier density Δn_{av} (as calculated from Eq. 2.6) is determined in the transient case via

$$\tau_{eff} = -\frac{\Delta n_{av}}{(d\Delta n_{av}/dt)}. \qquad (2.7)$$

This transient technique is only appropriate for the evaluation of photogenerated carrier lifetimes appreciably greater than the flash duration (> 100 μs).

On the contrary, in the QSS mode, during a long, exponentially decaying light pulse ($\sim$ 2 ms), the wafer conductivity is measured simultaneously with the illumination level using a calibrated light sensor. To convert the light intensity I_L [mWcm^{-2}] via the photon flux density ϕ_L [cm^{-2}s^{-1}]

$$\phi_L = \frac{\lambda}{hc} I_L \qquad (2.8)$$

into the generation rate G_L [cm^{-3}s^{-1}] (besides the wafer thickness W), the optical constant F [ ] has to be known, that is the fraction of incident light absorbed in the wafer under test:

$$G_L = \phi_L \frac{F}{W}. \qquad (2.9)$$

A bare wafer has an optical constant of about 0.7 when flat and 1 when textured for wafers with a standard thickness of 200 μm. A thicker wafer with an antireflection coating can exceed 1 due to the fact that an optical constant of 1 is equivalent to a photogenerated current of 38 mA/cm^2 at 1 sun in the Sinton QSSPC tool [SM06]. In practice, when measuring sufficiently high lifetimes (in order for the transient technique to be valid), the optical factor can be determined by seeking accordance between the curves acquired by the two measurement techniques. The effective lifetime τ_{eff} can then be calculated from Eq. 2.6 via the steady-state condition

$$\Delta n_{av} = G_L \times \tau_{eff}. \qquad (2.10)$$

The QSS technique only allows the measurement of lifetimes well below the flash decay constant, otherwise the quasi-steady-state-condition is no longer fulfilled.

2.1. Measurement techniques

In the generalized case there are both transient conditions and generation. τ_{eff} can then be rewritten as [NBA99]:

$$\tau_{eff} = \frac{\Delta n_{av}}{G_L - \frac{d\Delta n_{av}}{dt}}. \tag{2.11}$$

Eq. 2.11 reduces to the transient expression in Eq. 2.7 when $G_L = 0$ and to the QSS case in Eq. 2.10 when $d\Delta n_{av}/dt = 0$. The generalized analysis of the data thus allows the characterization of lifetimes over a wide range of values.

Several Sinton lifetime tester measurements with and without different neutral density (grey) filters (typically 30%, 15% and 5% transmission) and in the two measurement modes can be combined. In this way, the effective interface recombination rate can be evaluated over a very wide excess carrier density range, as shown in the injection level dependent lifetime measurement ($\tau_{eff}(\Delta n)$-curve) example of Fig. 2.5(a). From such multiple measurements, noisy data is removed and one single curve is displayed as shown in Fig. 2.5(b).

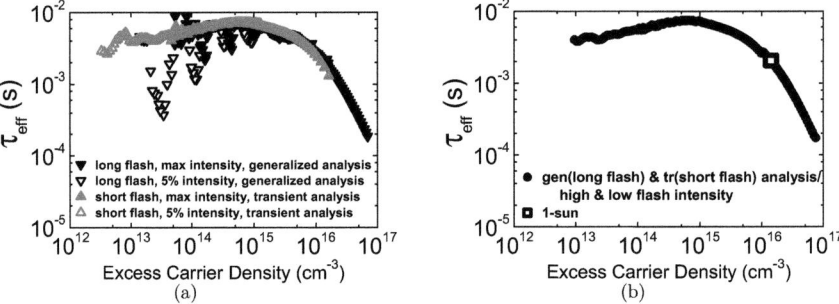

Figure 2.5: *a) Example of injection level dependent τ_{eff} as measured by the photoconductance decay technique with the Sinton lifetime tester in the transient (short flash) as well as in the generalized QSS mode (long flash), with and without grey filters. b) Wide injection level range $\tau_{eff}(\Delta n)$-curve combined from the multiple measurements shown in Fig. a) removing noisy measurement points. The square indicates τ_{eff} at Δn corresponding to the 1-sun illumination level on this sample (see text).*

The maximum open-circuit voltage (V_{OC} [V]) value of a slab of semiconducting material, but also of a final device is given by the splitting

2.1. Measurement techniques

of the quasi-Fermi levels ϕ_n and ϕ_p [V] ($n = n_0 + \Delta n = n_i e^{-q\phi_n/kT}$ and $p = p_0 + \Delta p = n_i e^{q\phi_p/kT}$ where n_0 and p_0 [cm^{-3}] are the electron and hole densities at thermal equilibrium):

$$V_{OC} = -(\phi_n - \phi_p) = \frac{kT}{q} \ln(\frac{np}{n_i^2}), \quad (2.12)$$

where n and p [cm^{-3}] are the total electron and hole densities and n_i [cm^{-3}] is the intrinsic carrier density. The Sinton lifetime tester simultaneously measures the flashlight intensity using a photodetector. Thus we can acquire both the lifetime and a prediction of the illumination level dependent V_{OC} (SunsV_{OC}) of a solar cell with this $\tau_{eff}(\Delta n)$-curve. Figure 2.6 shows a light-intensity dependent implied V_{OC}-curve calculated from

$$\text{impl}V_{OC} = \frac{kT}{q} \ln[\frac{(n_0 + \Delta n)(p_0 + \Delta p)}{n_i^2}]. \quad (2.13)$$

On the $\tau_{eff}(\Delta n)$-curve in Fig. 2.5(b) the $\Delta n, \tau_{eff}$-couple corresponding to the 1-sun illumination level is indicated. The diode ideality factor as a function of illumination can be determined by fitting as shown in Fig. 2.6 to $I_L = J_{0x} e^{qV_{OC}/(n_{0x}kT)}$, where J_{0x} [mAcm^{-2}] is the recombination current and n_{0x} [ ] is the ideality associated with it (more details about double-diode modeling are given in Sec. 3.3.3).

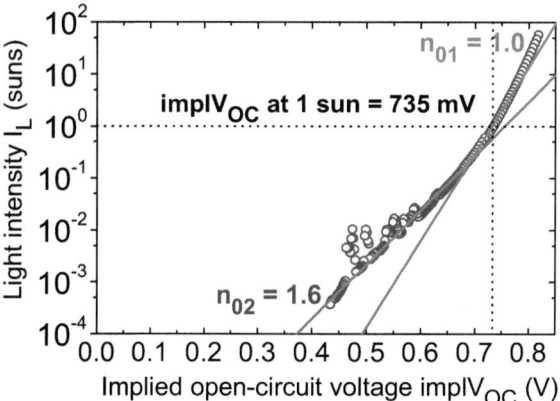

Figure 2.6: *Prediction of the illumination level dependent V_{OC} (implV_{OC} [V]) resulting from the measured $\tau_{eff}(\Delta n)$-curve in Fig. 2.5(b). The diode ideality factor is indicated for 1-sun as well as for low illumination.*

2.1. Measurement techniques

ILM lifetime mapping at ISFH While Sinton lifetime measurements rapidly give valuable information about τ_{eff} at varying illumination levels, the measured τ_{eff} is always an average over the sensing coil area (in our set-up 10 cm^2). Spatially resolved lifetime measurements are usually made by MW-PCD where a detector scans the wafer surface. The measurement times for such a scan are usually long and become longer as the measured lifetimes increase. The infrared camera lifetime mapping (ILM) technique offers fast lifetime mappings that become even faster when the measured lifetimes are high. Thus, when dealing with high lifetimes, ILM is a better choice for lifetime mapping than MW-PCD. At the Institut für Solarenergieforschung Hameln/Emmerthal (ISFH) ILM is performed in the emission mode [PB04]: the wafer is heated on a gold mirror to a temperature of 70 °C, while a diode array ($\lambda = 880$ nm) that is periodically turned on and off periodically photogenerates excess carriers in the wafer. These carriers modulate the infrared emission of the wafer by their change in density, through the phenomena of free carrier absorption. This periodic emission change is recorded with a lock-in technique by a long-wavelength infrared camera. The diode array's light intensity can be tuned to choose the illumination level and thus the photon flux density ϕ_L. The generation G_L that ϕ_L causes in a c-Si wafer is then given by its thickness W, while the optical factor F of the sample undergoing testing has to be known, as given by Eq. 2.9. All wafers used throughout this work are verified to have negligible transmission at the ILM's infrared excitation wavelength (absorption coefficient spectrum of c-Si in Fig. 2.2 and the transmission of thin textured wafers in Fig. 4.40). Additionally, a-Si:H and µc-Si:H thin-layers are transparent to infrared radiation (see again Fig. 2.2 and the absorption of a-Si:H/µc-Si:H layer stacks in Fig. 2.1(b)). As the infrared refractive indexes of amorphous and microcrystalline Si are similar to the one of c-Si, we adopt the optical factors of 0.7 for bare flat and 1 for textured c-Si for thin-film Si covered wafers.

The corresponding excess carrier density $\Delta n = \Delta n_{av}$ is calculated via the steady-state condition in Eq. 2.10 from the measured τ_{eff}. When comparing it to Sinton measurements, one has to pay attention to the fact that under the same light intensity, Δn varies according to the carrier lifetime.

2.1.2.2 Nanometrically resolved imaging of surfaces, interfaces and layer structures

The scanning electron microscope (SEM) makes an image of a sample's surface by scanning it with a high-energy beam of electrons in a raster scan pattern. In the most common imaging mode, the image of the sample surface is built by collecting low energy secondary electrons that originate from inelastic scattering within a few nanometers from the sample surface. A spatial resolution down to 1 nm is achievable in this image mode. To measure the thickness of a-Si:H layers on textured c-Si, a device cross-section is obtained by cleavage. The fracture surface yields enough contrast to identify the a-Si:H layer on the c-Si substrate. Samples were observed using a Philips XL30 Sirion microscope, a XL30 ESEM-FEG microscope and a Philips XL30 SFEG microscope. In general, SEM images are easier to interpret than TEM images (next paragraph).

2.1.2.3 Subnanometrically resolved imaging of interfaces and layer structures

High resolution transmission electron microscopy (HR-TEM) permits material characterization at the subnanometrical level. Therefore, TEM is a very powerful tool to gain information on the structure, phases and crystallography of materials. Its major drawback is that the sample preparation is time consuming and sample destructive. Here, we are mainly interested in examining the interfaces between the different materials composing a heterojunction, as these interfaces' quality is primordial for the fabrication of high performance devices.

We started to examine our samples at the IMT Neuchâtel [sam]. The samples were prepared using the cleaved corner method, and observed using a Philips CM200 microscope operated at 200 kV. The cleaved corner method consists of cleaving the thinned (about 150 µm) sample into about 1×1 mm^2 sized squares with a fine diamond stylus. It is thus simple and fast but the observable zone is small (just 2 corners of the square) and because of rapid sample contamination in the microscope, such cleaved corners can be observed only once.

In the framework of an EPFL [epf] semester work [DMP07], samples were then prepared by the tripod method. The tripod method consists first of gluing together two cleaved pieces of the sample. In this "sandwich", the layers to be studied face the glue. With the help of the tripod polisher, a wedge is produced whose thin side is thin enough to let electrons pass (Fig. 2.7). Thinning is finally accomplished with exposure to an ion beam, which simultaneously serves to clean the sample.

The main advantage of the tripod method sample preparation is the relatively large observable zone obtained. Principally, two different samples can be glued together, e.g. the front and the back thin-film Si layer stacks of a Si heterojunction solar cell. A drawback is the lengthy sample prepara-

2.1. Measurement techniques

tion. Also mechanical polishing as well as the ion beam thinning/cleaning can damage the sample.

A crystalline material interacts with the electron beam mostly by diffraction rather than absorption. In the medium resolution mode of the TEM, a bright field image is obtained by selecting the beams of transmitted electrons while blocking the beam of diffracted electrons. Regions having transmitted a large part of the electron beam thus appear clear in the contrast mode. In a dark field image, the transmitted electrons are blocked and the areas of the sample oriented in such a way as to efficiently diffract the incoming electron beam will appear clear, whereas a hole (only transmitting electrons) will appear dark. Finally, the highest spatial resolution of the TEM is obtained in the HR-TEM mode. Also known as phase contrast imaging, the images are formed due to differences in the phase of electron waves scattered through a very thin specimen. The high performing CM 300 UT-FE microscope of the EPFL operated at 200 kV allows for high-quality HR-TEM images.

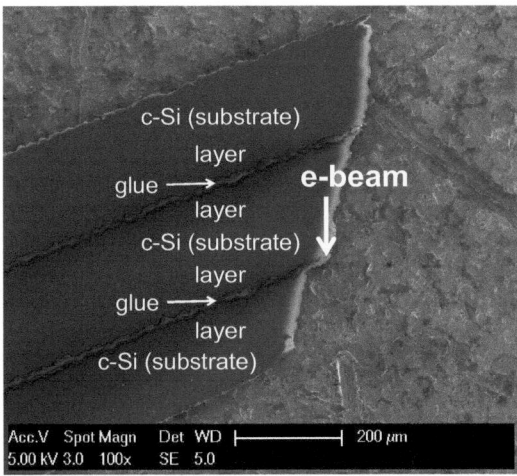

Figure 2.7: *SEM micrograph of a triple sandwich wedge prepared by the tripod method for TEM observation. The righthand side is sufficiently thin for TEM observations.*

In the framework of a continued informal collaboration with the EPFL, textured a-Si:H/c-Si heterojunction samples for HR-TEM observation could be prepared by the tripod method.

2.1.2.4 Solar cell efficiency measurement: current-voltage characteristic

The measurement of a solar cell's current-voltage characteristic (JV-curve) under standard test conditions permits the evaluation of a solar cell's maximum output power and thus its efficiency to convert sun light into electrical power. Standard test conditions consist of a temperature of 25 °C, a light intensity of 100 mW/cm^2 and a light spectrum close to the AM1.5g solar spectrum (shown in Fig. 2.8) like the one produced by the Wacom WXS-140S dual-lamp solar simulator used in this study.

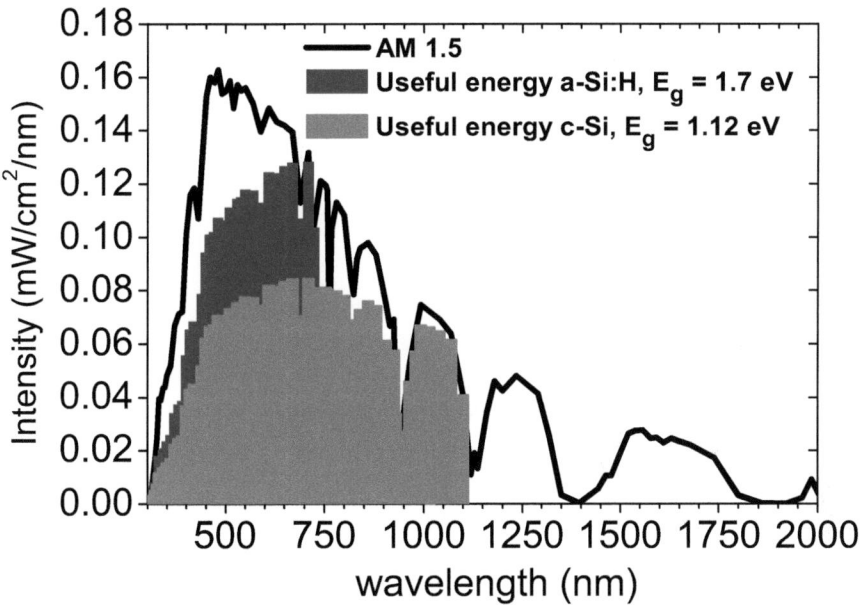

Figure 2.8: *The AM1.5g solar spectrum corresponds to the spectral distribution of sun light with a standardized intensity of 100 mW/cm^2 like the one measured on Earth with a sunlight incidence of 48° [GKV98]. The amount of incident solar energy that can be ideally converted into the output power of a solar cell depends on the bandgap of the material, as shown here for a-Si:H (bandgap 1.7 eV) and c-Si (bandgap 1.12 eV).*

Fig. 2.9 shows a typical current-voltage characteristic of a Si heterojunction solar cell indicating the values that can be determined from this plot.

2.1. Measurement techniques

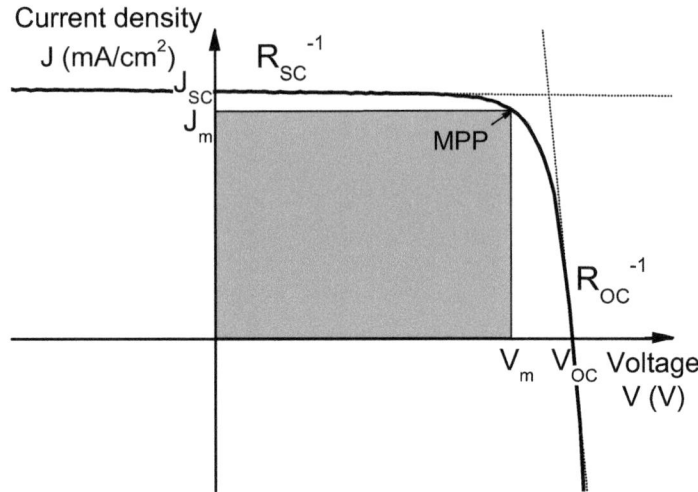

Figure 2.9: *Typical JV-curve of a Si heterojunction solar cell measured under standard AM1.5g illumination conditions.*

The values of V_{OC} [V] and J_{SC} [mAcm^{-2}] correspond to the open-circuit voltage and the short-circuit current density respectively. V_m [V] and J_m [mAcm^{-2}] denote the voltage and the current density couple where the power $P = V \times J$ is maximal, referred to as maximum power point (MPP). The fill factor (FF [%]) is the ratio of this actual maximum obtainable power to the theoretical (not actually obtainable) power:

$$FF = \frac{V_m \times J_m}{V_{OC} \times J_{SC}}. \quad (2.14)$$

The sunlight energy conversion efficiency η [%] denotes the maximal fraction of incident sunlight intensity, $I_L = 100$ mW/cm^2, that is delivered as power (P_m [mWcm^{-2}]) from the solar cell:

$$\eta = \frac{P_m}{I_L} = \frac{FF \times V_{OC} \times J_{SC}}{I_L}. \quad (2.15)$$

The slope of the JV-characteristic at the V_{OC} and J_{SC} points are given by $R_{OC} = (\partial J/\partial V)^{-1}_{J=0}$ [mΩcm^2] and $R_{SC} = (\partial J/\partial V)^{-1}_{V=0}$ [mΩcm^2]. In a first approximation, they are close to the values of the series resistance R_s and the parallel resistance R_p that a simplified solar cell equivalent circuit considers.

2.1. Measurement techniques

Note that the set-up measures the total current and not the current density. The surface area is typically 0.2 cm², but it is not always exactly the same as shown in Fig. 2.20. To avoid uncertainties in the determination of the absolute value of J_{SC}, it is given by integrating the wavelength dependent current delivered from a small spot of a cell, i.e. from the EQE measurement discussed in Sec. 2.1.2.5.

While such standard test conditions serve to compare different solar cells, for practical applications, the solar cell characteristics' behavior at lower illumination intensities and at higher operating temperatures is crucial.

2.1.2.5 External quantum efficiency

The external quantum efficiency (EQE [%]) measures the ratio of the number of electrons flowing into the contact to the number of incident photons (measured at 0 V except when stated otherwise). Therefore, the EQE measures the light absorption probability within the active device thickness times the probability of the light-generated carriers to reach the outer contacts. EQE measurements between 350 and 1100 nm are performed in this work. For this, the solar cell is illuminated with a small beam spot of a well defined size (about 1×3 mm²). The current delivered from this spot on the cell at a given wavelength is measured and divided by the incident light intensity I_L that is measured with a reference detector, whose quantum efficiency is known. Figure 2.10(a) shows the EQE measurement of a flat Si heterojunction solar cell.

This EQE is the product of the internal quantum efficiency (IQE [%]) and optical losses due to the solar cell's total external reflection R_{cell} assuming that there are no transmission losses, i.e. $T_{cell} = 0$ and thus $A_{cell} = 100 - R_{cell}$:

$$EQE = IQE(100 - R_{cell})/100. \qquad (2.16)$$

The IQE is thus the probability of a photon to enter the solar cell and to yield an electron in the external solar cell current circuit. Comparing solar cells' IQEs permits to abstract from front surface reflection, which is e.g. useful when comparing a flat and a textured solar cell consisting of the same layers. Parasitic absorption in the electrically inactive ITO, the p-layer and the back contact is still contained in IQE. In a first approximation, carriers photogenerated in the doped a-Si:H/μc-Si:H emitter layer cannot be collected due to the doped layers high defect density. (However, from

2.1. Measurement techniques

the numerical simulation of amorphous Si solar cells there are hints that the doped layers' photoelectrical activity is higher than generally thought.) IQE* [%] makes abstraction of the ITO's parasitic absorption A_{ITO} [%]:

$$IQE^* = \frac{100 \times EQE}{(100 - R_{cell}) - A_{ITO}}. \tag{2.17}$$

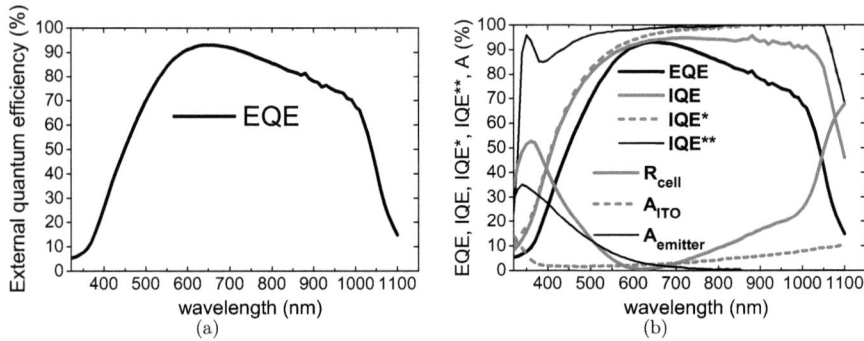

Figure 2.10: a) External quantum efficiency (EQE) measurement of a Si heterojunction solar cell. b) Internal quantum efficiencies IQE, IQE* and IQE** as calculated from the EQE measured in Fig. a) together with the measurement of the cell's reflectance and the absorption of the ITO (Fig. 2.1(a)) and emitter layer stack deposited on glass (Fig. 2.1(b)). For details, see text. Measurement errors of ±2% have to be expected in the absorption related measurements and an additional measurement error comes from the different spot sizes in EQE and photospectrometry measurements.

Finally IQE** [%] is the probability of a photon to be absorbed in the c-Si wafer and to yield an electron in the external solar cell current without recombining beforehand. It can be calculated if the solar cell's total reflection R_{cell}, the absorption in the ITO A_{ITO} and the a-Si:H/µc-Si:H emitter layer $A_{emitter}$ [%] is known (neglecting the absorption in the back contact):

$$IQE^{**} = \frac{100 \times EQE}{(100 - R_{cell}) - A_{ITO} - A_{emitter}}. \tag{2.18}$$

Fig. 2.10(b) shows IQE, IQE* and IQE** as calculated from the measurements of the solar cell EQE in Fig. 2.10(a), this solar cell's total reflection

2.1. Measurement techniques

and the corresponding ITO and emitter layer absorption measurements on glass described in Sec. 2.1.1.2. Because of the wavelength-dependance of the absorption coefficient (Fig. 2.2), the shorter wavelength photons photogenerate free carriers in the c-Si bulk closer to the front surface. These are thus more affected by front surface recombination. Similarly, the longer wavelength photons are more affected by back surface recombination. They are generated the nearer to the back surface, the thinner the c-Si wafer is. Comparing solar cell's IQE**s allows the abstraction of the solar cell properties from the ITO characteristics and the thickness of the a-Si:H/µc-Si:H emitter layer stack. It also gives a direct indication of front and back surface passivation quality.

Note that measurement errors can accumulate up to 4% when passing from EQE over IQE and IQE* to IQE**. This is because of small thickness non-uniformities that strongly influence the result and because of the differences in the illuminating spot sizes when measuring EQE and R_{cell}.

The short-circuit current density that is reached by a given QE curve can in turn be calculated by

$$J_{SC} = q \int_0^\infty \phi_{AM1.5g}(\lambda) QE(\lambda) d\lambda, \qquad (2.19)$$

where $\phi_{AM1.5g}(\lambda)$ [cm^{-2}s^{-1}] is the photon flux related to the spectral AM1.5g sunlight intensity $I(\lambda)$ (Fig. 2.8) by $\phi_{AM1.5g}(\lambda) = I_{AM1.5g}(\lambda) \times \frac{\lambda}{hc}$ (Eq. 2.8).

2.1.2.6 Acquiring series-resistance-less JV-curves: SunsV_{OC} measurements

The SunsV_{OC} setup on the WCT-100 Sinton Consulting equipment measures the solar cell V_{OC} as a function of the light intensity that is monitored by a photodiode, as shown in Fig. 2.11(a). Using the superposition principle, the SunsV_{OC} data can be represented in the familiar JV format, shown in Fig. 2.11(b) by triangles. For this, an implied terminal current (J_{impl} [mAcm^{-2}]) is determined for each V_{OC} from the normalized light intensity $I_L/I_L(1\text{ sun})$ and from an estimated J_{SC} under 1-sun illumination. Note that I_L, the incident light intensity penetrating the sample, is assumed to be proportional to the luminous intensity with these considerations:

$$J_{impl}(V_{OC}) = J_{SC}(1 - \frac{I_L(V_{OC})}{I_L(1\text{ sun})}). \qquad (2.20)$$

2.2. Pre-deposition wafer treatment

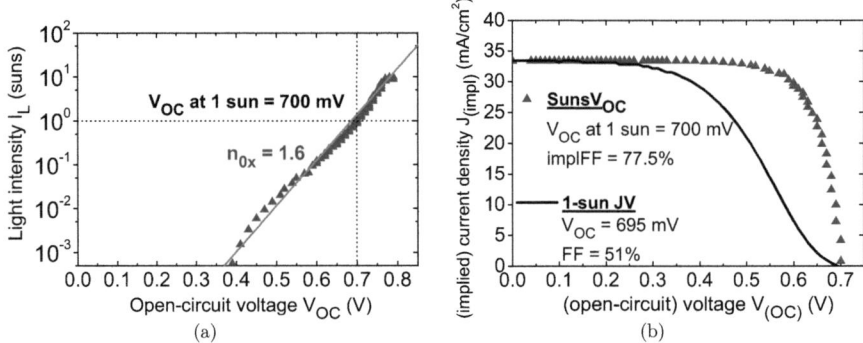

Figure 2.11: *a) SunsV$_{OC}$ measurement and b) implied JV-curve calculated from it (triangles), compared to the actual 1-sun JV-curve of the same cell, but drawing current from it (line).*

In such a configuration, the series resistance (R_S) has no effect on V_{OC} since no current is drawn from the cell. The resulting implied JV-curve is thus the JV-curve of the studied solar cell without series resistance losses. Hence, the effects of shunt and series resistance on the final device performance can be easily elucidated. The Sinton SunsV$_{OC}$ tool uses the same flash lamp as in lifetime testing, consequently SunsV$_{OC}$ curves can be taken very quickly and simply.

The respective contribution of Si based thin-layers and TCO resistances as well as the one of blocking junctions to the actual 1-sun JV-curve (Sec. 2.1.2.4, shown by a line in Fig. 2.11(b)), remains indiscernible in the SunsV$_{OC}$ curve. SunsV$_{OC}$ measurements can only immediately detect FF limitations due to unfavorable injection level dependent surface recombination velocities as shown in Fig. 2.11(b), illustrating the example of a Si heterojunction solar cell based on lightly n-type doped c-Si.

2.2 Pre-deposition wafer treatment

The performance of an a-Si:H/c-Si heterostructure is first of all determined by the c-Si surface's condition before a-Si:H layer growth. As a first condition, the c-Si surface has to be clean enough to be ready for passivation after a simple HF-dip for native oxide removal. Contrary to high-temperature passivation such as silicon dioxide (SiO_2), no cleaning of as-purchased polished c-Si wafers is necessary, probably because the low process temperatures used are insufficient to activate the recombination activity of remaining surface contaminants. However (in contrast to flat wafer pre-deposition cleaning), after KOH-texturing, a cleaning step is indispensable to remove remaining contaminants.

As a second condition, reoxidation after native oxide removal has to be prevented for best interface qualities. In contrast to SiO_2 passivation where the top c-Si surface is consumed during surface oxidation, the a-Si:H/c-Si heterointerface lies exactly at the top c-Si surface. Due to the major influence of this Si heterointerface, even minor variations in the c-Si surface conditions are a source for process instabilities.

2.2.1 Native oxide removal

In ambient air, the bare c-Si surface oxidizes, rapidly forming a native oxide of a few nm thickness. Contrary to a thermally grown oxide, this native oxide has a defective interface with c-Si and therefore yields no surface passivation. Additionally, the native oxide is a (superior) electric insulator that hinders current transport in a-Si:H/c-Si heterojunction solar cells. In addition, the growth of thin-film Si layers crucially depends on the microstructure of the underlying substrate. Under the same processing conditions, a layer can grow epitaxially on c-Si yet amorphous on a thin native oxide on c-Si [TFM+95].

Hydrofluoric acid (HF) efficiently removes the native oxide from the c-Si surface [Ker76]. As the native oxide is a low-quality oxide, it is rapidly etched away by highly diluted HF. As the HF etch selectivity of oxide over silicon is extremely high, the silicon surface roughens only after prolonged immersion times (since water will slowly oxidize the surface of the silicon and HF will etch this oxide away) [HK67]. In addition to its quality of not being critically dependent on the process parameters, the HF-dip itself is a process that is easy to monitor. This is because the presence of a surface oxide manifests itself by a hydrophilic surface, contrary to the hydrogen-covered c-Si surface left after the completed HF-dip, which is hydrophobic. We chose a HF dilution of 4% in deionized (DI) water and a dip-time of 45 seconds to ensure a complete native oxide removal, although the wafer surfaces becomes hydrophobic already after a few seconds. For all double side polished wafers that were purchased, the native oxide removal by diluted HF is sufficient as a cleaning step. This is probably because at our low processing temperatures (< 200 °C), remaining surface contaminants are not sufficiently activated to be detrimental for surface passivation.

Wafers immersed in a HF solution yield some of the highest lifetimes,

2.2. Pre-deposition wafer treatment

making them among the best surface passivations ever reported [YAC+86, LC88]. These are plotted in Fig. 2.12(a) taking data from Yablonovitch et al. [YAC+86]. But the passivation provided is not stable with time, as indicated by the rapid lifetime decrease with time after a HF-dip in ambient atmosphere, shown in Fig. 2.12(b).

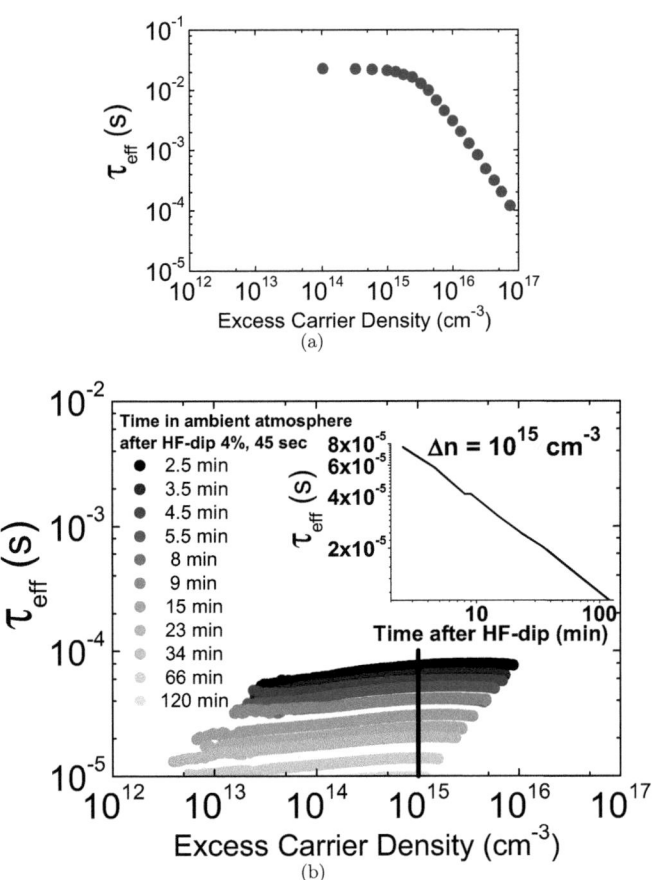

Figure 2.12: *a) 150 Ωcm n-type doped <111> orientated 250 μm thick c-Si wafer immersed in HF, yielding among the highest ever measured lifetimes [YAC+86], but b) rapid lifetime decrease in ambient air after HF-dip due to reoxidation on 3 Ωcm n-type doped <100> orientated 300 μm thick c-Si (bulk lifetime = 1.2 ms).*

2.2. Pre-deposition wafer treatment

For a-Si:H/c-Si heterojunction solar cells, the c-Si surface is directly at the *pn*-junction, making a native oxide free c-Si surface crucial. That is why we only very shortly shower-rinse a HF-dipped wafer with DI water (if at all). Then the wafer is loaded as fast as possible into the load-lock of the deposition chamber and finally the pump-down and substrate heating times before the deposition are minimized.

From minor variations in the duration of the wafers' stay in ambient atmosphere after the HF-dip one must expect variations in the final Si heterostructure performances (Fig. 2.12(b)). In view of solar cell inline processing and process reproducibility, it would be desirable to replace the wet chemical HF-dip by a plasma process. In the best case scenario this plasma etching would have the same selectivity of SiO_x over Si as HF, and not only yield an initial excellent surface passivation but also a stable one. Initial tests of hydrogen (H_2)-plasma etching in the VHF-PECVD deposition chamber showed that the native oxide can be etched away but also that this process is critical because over-etching damages the electronic properties of the c-Si surface. The native oxide does not have a uniform composition and thickness. Therefore, H_2-plasma etching performed with the process parameters used is not a suitable alternative to the conventional HF-dip. Later on, commercial plasma cleaning systems to free microelectronics metal strips from native oxides before wire bonding were used. But removing the native oxide from the c-Si surface without destroying the latter's electronic properties by plasma processes proved to be difficult.

2.2.2 Textured c-Si cleaning issue

Light-trapping schemes have to be applied to reduce the solar cell current losses by reflection. They also increase the optical path of photons entering a solar cell, in turn increasing the latter's absorption probability. In the case of monocrystalline Si, random pyramidally textured surfaces effectuate efficient geometrical light-trapping. These pyramids are formed by profiting from the KOH etch selectivity of Si<100> over Si<111> planes [Bea78]. The pyramid size and distribution critically depends on the initial wafer surface, the process temperature and the KOH and IPA concentrations used [SSB+03]. Textured wafers cannot be purchased from a wafer company and other institute's texture is optimized for other solar cell applications. The company Solarworld [sol] kindly textured wafers for us with the pyramids that we requested. Whereas for polished purchased wafers an additional cleaning is not necessary, after texturing, an additional cleaning step proved to be required. Cleaning of our textured wafers was done with the standard cleaning procedure at the "Sensors, Actuators and Microsystems Laboratory" of the IMT Neuchâtel [sam], involving the following process steps:

1. H_2SO_4 + peroxide; 120 °C; 10 minutes
   $\rightarrow$ Removing organic residues on top of the native oxide layer

2. Rinsing in deionized water

3. BHF (7:1); 20 °C; 1 minute
   $\rightarrow$ Removing the native oxide layer (hydrophobic surface)

4. Rinsing in deionized water

5. HNO_3 70%; 115 °C; 10 minutes
   $\rightarrow$ Homogeneous thin oxide layer (hydrophilic surface)

6. Rinsing in deionized water

Thanks to the thin protective oxide layer left behind by this cleaning, the cleaned wafers can be stored (preferably in nitrogen) before giving them a final HF-dip prior to processing. This is again probably due to our low processing temperatures that do not effectively activate a small amount of contaminants that would already be detrimental in high-temperature processes.

Most likely, this microelectronic cleaning sequence is too sophisticated for our application and would be too costly for solar cell production. That is why in the future simpler cleanings with fewer steps need to be studied. An inline plasma cleaning would again be ideal from the point of view of an industrial implementation.

2.3 VHF-PECVD deposition of amorphous and microcrystalline silicon

There are several methods for the deposition of thin-film silicon layers such as amorphous and microcrystalline Si. Only some of them yield a sufficiently high layer quality for the use in optoelectronic applications. Hot wire chemical vapor deposition (HWCVD) gives good layer properties [MNSC91, BTY+08, KFC+01, KFC+03] but has disadvantages for industrial use. Actually, the most widespread deposition technique consists of the decomposition of gases by a plasma, called plasma enhanced chemical vapor

2.3. VHF-PECVD deposition of amorphous and microcrystalline silicon

deposition (PECVD). Silicon thin-film depositions by PECVD most often use 13.56 MHz as the plasma excitation frequency. The use of plasma excitation frequencies in the very high frequency (VHF) range of 40 to 135 MHz leads to an increase of the deposition rate without losing in material quality, thanks to a reduced ion bombardment [HDH+92]. We will see that the use of VHF-PECVD also governs some specific advantages over other deposition techniques when growing amorphous silicon on crystalline silicon. VHF-PECVD was pioneered by our research group [CWFS87] and due to its successful industrial implementation by Oerlikon Solar [oer] and Flexcell [fle], it is still the principal technique currently used in our lab.

2.3.1 VHF-PECVD deposition chamber

Amorphous and microcrystalline silicon layers are grown by very high frequency plasma enhanced chemical vapor deposition (VHF-PECVD). For this work, the single chamber deposition system shown in Fig. 2.13 was used. It includes a load-lock that permits keeping a low base pressure of 1×10^{-8} to 1×10^{-7} mbar. It is a parallel plate deposition system with two round electrodes, separated by 14 mm, having surfaces of 133 cm^2. The upper electrode is the grounded electrode and contains the substrate holder. The lower electrode is fed by a VHF generator at a frequency of 70 MHz where a match box (two capacitances that can be adjusted) ensures maximal power injection into the plasma. Both electrodes are heated up to 130 - 350 °C. The process gas flux is injected from the side walls of the lower electrode while a butterfly valve is used for the regulation of the process pressure. Finally, the plasma is lit by inducing an electrostatic discharge by means of a piezoelectric lighter.

Single chamber reactors without plasma chamber cleaning capability are not optimal in view of doping cross-contamination. The large amount of dummy layers that need to be deposited after each doped layer deposition reduce the reproducibility. Furthermore, the use of a plasma confinement box [PSH+00] and a smaller deposition chamber volume as compared to the electrode surface would help better defining the plasma deposition parameters. The optimized process parameters obtained with this simply configured small area single chamber deposition system can be then transferred to an industrial-like large area VHF-PECVD deposition system. The specifically used system is working at a plasma excitation frequency of 40 MHz. An Oerlikon plasma box [PSH+00] and plasma chamber cleaning capability both greatly improve the process reproducibility.

2.3. VHF-PECVD deposition of amorphous and microcrystalline silicon

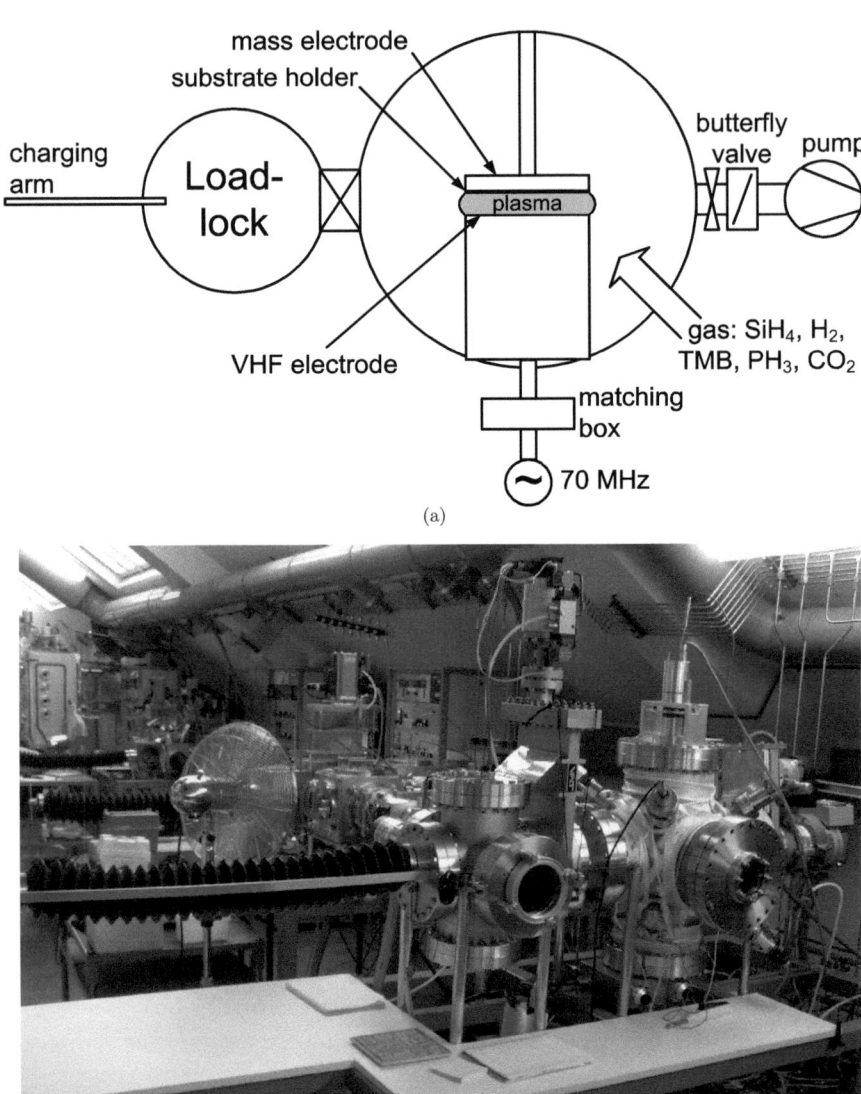

Figure 2.13: *a) Sketch and b) photo of the VHF-PECVD deposition system used for this work.*

2.3. VHF-PECVD deposition of amorphous and microcrystalline silicon

2.3.2 Intrinsic amorphous silicon

The quality of intrinsic (i) a-Si:H layers strongly depends on the process parameters used. Adding hydrogen (H$_2$) to the silane (SiH$_4$) gas flux reduces the a-Si:H's defect density [Pla99]. While layers deposited at very low process temperatures (< 100 °C) have a high defect density, best device-quality layers are deposited at higher temperatures (> 130 °C) by seeking a dense structure. The intrinsic a-Si:H layer used for c-Si surface passivation, when not otherwise mentioned, was optimized as an absorber layer for a-Si:H solar cell application, where the defect density is crucial. But in fact, the properties of these typically 300 nm thick layers cannot be correlated to those of a layer growing within the same process parameters only up to a thickness of 5 nm. This standard layer's deposition parameters are a hydrogen dilution $H_{dil} = \frac{[H_2]}{[SiH_4]}$ of 2.7 (corresponding to a silane concentration $SC = \frac{[SiH_4]}{[SiH_4]+[H_2]}$ of 27%) and a deposition temperature of 200 °C. Table 2.1 summarizes all deposition parameters of the i a-Si:H layers used for c-Si surface passivation and a-Si:H/c-Si heterojunction formation in this study. The growth rate of the i a-Si:H layers is constant, which means that while a 5 nm thick layer grows in 15 seconds a 250 nm thick layer does so in 50 × 15 seconds.

Layer	Silane conc. (%)	Deposition temp (°C)	Pressure (mbar)	Power density (mW/cm^2)	Growth rate (Å/s)
i a-Si:H std	27	200	0.4	19	3.1
i a-Si:H high dil	10	200	0.2	19	1.9
i a-Si:H no dil	100	200	0.3	19	4.3

Table 2.1: *Process parameters of the i a-Si:H layers used in this study for c-Si surface passivation and a-Si:H/c-Si heterojunction formation.*

2.3.3 (Intrinsic) amorphous silicon microdoping

The average charge on the silicon bonds can be varied without much influencing the defect density by adding only a few ppm (parts per million) of doping gas to the gas mixture used for the i-layer growth (phosphine (PH$_3$) for n-type doping and trimethylborane (TMB) for p-type doping, both diluted in H$_2$) [Str85]. Such low-level doping is called microdoping. Heavier doping results in an increase in the defect density and does, thus, no longer permit the variation of the charge on the defects independently of their density. The high defect density in doped thin-film layers is also the reason why their direct deposition on c-Si leads to increased interface recombination and thus to lower solar cell open-circuit voltages. To achieve a controlled microdoping, the dopant gases that are at our disposal at a concentration of 1000 ppm (PH$_3$) and 500 ppm (TMB) in hydrogen would need to be prediluted in a mixing chamber to the 10 ppm range. Having no such mixing chamber at disposal, we simply add the smallest possible dopant gas flux to the standard i-layer gas mixture (process parameters in Tab. 2.1), that result in a 10 ppm phosphorous doped n-layer and a 5 ppm boron doped p-layer. At these low concentrations, the dopant incorporation efficiency is around one [Fis94].

2.3.4 Doped microcrystalline silicon

Doped thin-film Si layers are used only when stacked with i a-Si:H layers, because (as discussed in Sec. 2.3.3) the direct deposition on the c-Si wafers of the more defective doped layers (i.e. without intermediate i-layers) leads to increased interface recombination.

For best carrier extraction and highest built-in potential achievement we intend to make the doped thin-film Si layers as conductive as possible. While the best p a-Si:H layers only achieve conductivities of 1×10^{-4} S/cm, the best conductivities ($>$ 1 S/cm corresponding to activation energies lower than 20 meV) are achieved by doped Si material deposited near the amorphous/microcrystalline (a-Si:H/µc-Si:H) transition. The reason for this is that doping is more efficient in µc-Si:H than in a-Si:H, but amorphizes the layer after a certain optimum dopant concentration point, i.e. a further increase of the dopant concentration amorphizes the layer too much and decreases its conductivity again [Pra91]. In addition, it is easier to obtain an ohmic contact between this highly doped p-type a-Si:H/µc-Si:H transition layer and ITO.

In the initial growth stage of layers grown with deposition conditions close to the a-Si:H/µc-Si:H transition, a fully amorphous incubation layer has frequently been observed. The emitter p (or n) a-Si:H/µc-Si:H transition layer has to be as thin as possible because carriers photogenerated within it almost do not contribute to the solar cell's current generation but are lost by absorption. Additionally, the growth of µc-Si:H material is highly substrate dependent [KTMI96, VSBM$^+$05], e.g. a layer grown within the same deposition parameters can become microcrystalline when deposited on glass but completely amorphous when deposited on another a-Si:H layer. The deposition parameter space is wide for growing thick highly conductive doped thin-film Si layers on glass, but becomes narrow when the same layer conductivity is requested for an ultra-thin doped layer

2.3. VHF-PECVD deposition of amorphous and microcrystalline silicon

grown on an a-Si:H sub-layer [Flu95,Per01]. This parameter space becomes even more narrow because in such small area PECVD systems, the power/pressure parameter couple has to be chosen such as to give the most uniform deposition. In analogy to the end-targeted configuration for use in a-Si:H/c-Si HJ solar cells (c-Si / 5 nm i a-Si:H / 15 nm doped a-Si:H/µc-Si:H), we only directly develop thin doped a-Si:H/µc-Si:H transition layers on 5 nm i a-Si:H on glass.

Also for the deposition of amorphous solar cells in the *nip* configuration, an ultra-thin *p*-layer has to grow on a-Si:H. It was shown by Pernet *et al.* [PHH+00], that the application of interface treatments and delayed interface treatments is crucial for success. The layer conductivity is first optimized simply by four point probe measurements after a standard anneal of 90 min at 180 °C under nitrogen(N_2)-atmosphere. Raman crystallinity measurements greatly help to find out if a specific layer is not conductive enough because it is too microcrystalline, too amorphous or just not sufficiently doped. Raman crystallinity measurements of the i a-Si:H / doped a-Si:H/µc-Si:H layer stacks do not allow the evaluation of the crystallinity of the doped layer in absolute terms but permit a comparison of the samples between each other. Once a layer is judged to be good, the dark conductivity measurement is made to more precisely evaluate its conductivity and determine its activation energy. These measurements can also be used to track the VHF-PECVD chamber's stability over time. The deposition parameters in Tab. 2.2 were combined together with i) an intercaled CO_2-plasma, ii) a final H_2-plasma treatment for *p*-layer deposition, and iii) a H_2-plasma treatment on the i a-Si:H layer before *n*-layer deposition. This led to the good doped layer characteristics summarized in Tab. 2.3.

Layer	Silane conc (%)	Doping conc (%)	Deposition temp (°C)	Pressure (mbar)	Power dens (mW/cm^2)	Growth rate (Å/s)
p a-Si:H/µc-Si:H	0.75	1.42	200	0.6	53	0.37
n a-Si:H/µc-Si:H	0.94	0.99	200	0.9	45	0.37
n_2 a-Si:H/µc-Si:H	0.94	2.01	200	0.9	45	0.33

Table 2.2: *Process parameters of the doped thin-film Si layers growing near the a-Si:H/µc-Si:H transition used in stack with i a-Si:H as emitter and BSF layers in Si heterojunction solar cells.*

As can be seen from Tab. 2.3, *n*-type doped a-Si:H/µc-Si:H layers can be made much more conductive than their *p*-type doped counterparts. Also fully amorphous *n* a-Si:H layers would provide the requested ohmic contact to ITO. Besides, the use of doped a-Si:H layers would be preferable in terms of deposition parameter stability and tolerance. But we only dispose of PH$_3$ diluted to 1000 ppm in H$_2$ as a dopant gas. Increasing the doping concentration means simultaneously decreasing much the silane concentration so that optimal process parameters cannot be used. Note that although these doped a-Si:H/μc-Si:H layers have to be grown thicker than their completely amorphous counterparts, they have a lower short-wavelength absorption coefficient due to their smaller bandgap (verified in absorption spectrum of thin-film Si in Fig. 2.2) and they are probably also photoelectrically more active.

Layer stack	Doped layer thickness (nm)	Conductivity (S/cm)	Activation energy (eV)
p a-Si:H/μc-Si:H on 5 nm i a-Si:H	20	0.2	0.04
n a-Si:H/μc-Si:H on 3 nm i a-Si:H	22	32	0.02

Table 2.3: *Thickness, conductivity and activation energy of the doped a-Si:H/μc-Si:H layers grown with the deposition parameters of Tab. 2.2 on i a-Si:H on glass (standard i-layer deposition parameters from Tab. 2.1).*

2.4 Contact formation

The sheet resistance of the doped thin-film Si layers forming the Si HJ's emitter and BSF is superior (~ 10 K $- 1$ MΩ/□) to that of their diffused c-Si counterparts ($50 - 120$ Ω/□). Contacting Si HJ solar cells the same as standard crystalline Si solar cells by mm-spaced metal fingers would thus result in high series resistance losses, much reducing FF. The inclusion of a transparent conductive oxide (TCO) layer before the metal grid deposition greatly improves current collection by minimizing series resistances. The requirements for this TCO are high, as it has to i) simultaneously form a lossless contact to doped thin-film Si and metal, ii) be as transparent as possible over a wide wavelength range, and iii) act as an antireflection coating by index-matching between Si and air (finally Si and encapsulant in a module). In this study, we directly contact the TCO layer for solar cell property measurements, but as resistive losses rapidly become high without further metal grid formation, we have to limit the size of such gridless cells. Practically, we structure the TCO's surface into $(4.5 \text{ mm})^2$ sized cells, just large enough for EQE and JV measurements but sufficiently small to prevent FF-losses due to the TCO's resistivity.

On the rear side of the full c-Si wafer, a metal layer back contact (Al or Ag) simultaneously serves as a mirror for enhanced infrared absorption in the c-Si bulk. A TCO layer is inserted before the metal deposition

2.4. Contact formation

to protect the thin-film Si from metal diffusion and serves as an optical index-matching layer.

2.4.1 Transparent conducting contact deposition: ITO

Similarly to amorphous thin-film Si solar cells, a transparent conductive oxide (TCO) serves as an ohmic contact to transport photogenerated charge carriers with less resistive losses to the metal contacts. This TCO serves at the same time as an antireflection coating by index-matching between Si and air, i.e. Si and encapsulant in a final module. To simultaneously reach a low sheet resistance, a maximal optical transparency and an antireflection behavior, Indium Tin Oxide (ITO, $In_2O_3{:}SnO_2$) is here the material of choice.

In this study, ITO for HJ solar cells is deposited by DC sputtering in the MRC system 603. As substrate heating during ITO deposition is not possible, the sheet resistance of initial ITO layers is high. A standard annealing at 180 °C during 90 min under nitrogen atmosphere almost triples the ITO layers conductivities measured by four point probe. When optimizing ITO layers, one is always confronted by having to find a compromise between highly conductive and highly transparent layers. The reason for this is that the free carriers that are mainly responsible for a low sheet resistance absorb infrared light, as can be seen from the ITO absorption spectra in Fig. 2.14. Generally, a TCO's conductivity σ $[(\Omega cm)^{-1}]$ is given by:

$$\sigma = N \times \mu \times q, \qquad (2.21)$$

where N $[cm^{-3}]$ is the free carrier density and μ $[cm^2V^{-1}s^{-1}]$ the free carrier mobility. To minimize free carrier absorption, a high conductivity is thus favorably due to a high free carrier mobility instead of a high free carrier density. The resistivity ρ $[\Omega cm]$ is just the inverse of the conductivity, $\rho = \frac{1}{\sigma}$, and independent of the TCO thickness. The sheet resistance $R_{sq} = \frac{\rho}{d}$ $[\Omega/\square]$ (where d is the TCO layer thickness) is then the decisive parameter for the series resistance of a solar cell device. The thicker the TCO layer, the lower R_{sq} is, but also the higher the infrared absorption due to free carriers. In our case, when choosing the ITO thickness, one has to consider that the ITO layer should simultaneously act as an antireflective coating. That is why its refractive index n_r [ ] chooses the ITO layer's thickness to be about 85 nm.

Due to its old electronics, this MRC system suffers from frequent breakdowns after which standard process parameters had to be readjusted twice during this study, as shown in Tab. 2.4, including an exchange of the oxy-

2.4. Contact formation

gen/argon (O_2/Ar) gas bottle. The corresponding electronical properties of the resulting ITO layers deposited on glass are listed in Tab. 2.5 and their optical absorption is shown in Fig. 2.14.

name	O_2/Ar (%)	Pressure (μbar)	Power density (W/cm^2)	Deposition rate (nm/s)
ITO$_1$	3	11	1	1.1
ITO$_2$	2	17	1	1.0
ITO$_3$	2	9	1	0.9

Table 2.4: *Deposition parameters of standard ITO layers used in this study, all deposited at room temperature.*

name	Resistivity (Ωcm)	Sheet resistance (Ω/sq)	Free carrier density (cm^{-3})	Free carrier mobility (cm^2/(Vs))
ITO$_1$	2.2×10^{-4}	27	1.4×10^{21}	20
ITO$_2$	2.5×10^{-4}	31	9.5×10^{20}	24
ITO$_3$	2.7×10^{-4}	31		

Table 2.5: *Electrical resistivity and sheet resistance as resulting from free carrier density and mobility of ITO layers deposited up to a thickness of 85 nm on glass with the process parameters listed in Tab. 2.4.*

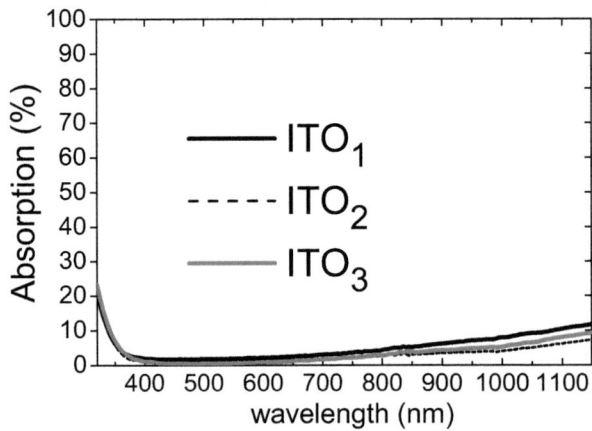

Figure 2.14: *Optical absorption of 85 nm thick ITO layers deposited on glass.*

2.4. Contact formation

In summary, for Si HJ solar cell fabrication, the low sheet resistance of the ITO is important for minimizing series resistance losses before the current is collected by the metal fingers for maximizing the FF. On the other hand, a high transparency and an appropriate refractive index are important for maximizing the photogeneration in the c-Si and thus maximizing J_{SC}. But there is a third parameter that is crucial for a properly functioning Si HJ solar cell: the ITO's work function ϕ_{ITO} [V], that needs to be at least of the same value as the work function of the p a-Si:H layer $\phi_{p\,a-Si:H}$ [V] to form a favorable contact (see Fig. 5.4 for definitions). ϕ_{ITO} is hardly experimentally accessible and varies from 4.3 eV to 5.1 eV as has been investigated for organic light emitting diode (OLED) fabrication [SIOS00, KLM+00, MHS+00]. ϕ_{ITO} is thus lower than the generally assumed value of $\phi_{p\,a-Si:H}$. This can lead to severe current transport deteriorations manifesting itself in S-shaped JV-curves. Using the much more conductive p-type doped a-Si:H/µc-Si:H transition material layers for Si HJ solar cell formation, an ohmic contact between the thin-film p-type Si and the ITO that does not deteriorate the JV-curve shape can be achieved more easily.

ITO deposition with this MRC system is very fast and generally reproducible. But to be able to optimize the ITO's work function and to find the best compromise between a low sheet resistance and a high transparency, substrate heating as well as tunability of the oxygen process gas content in argon would be required.

While ITO with a refractive index n_r deposited at a thickness d effectively minimizes the high bare flat c-Si substrates reflectivity around a wavelength $\lambda(n_r, d)$, the reflectivity at other wavelengths is quite high. One aims to primarily minimize reflectivity around a wavelength of 650 nm, where a maximal amount of incident solar energy can be converted into the output power of a c-Si solar cell (see Fig. 2.8). Figure 2.15 shows how ITO_1 having a refractive index of 1.9 around a wavelength of 650 nm (inset in Fig. 2.15 [DMP07]) minimizes the flat c-Si solar cells reflection losses when deposited at a thickness of 85 nm (destructive interference in a quarter-wavelength thick layer, i.e. $\frac{1}{4} \times \frac{\lambda}{n} = d$).

Fig. 2.15 indicates that a lot of light is prevented from entering the c-Si bulk for a flat wafer. When texturing the c-Si surface with random pyramids, the bare Si surface's overall primary reflectivity $Refl(\lambda)$ is reduced to $Refl^2$ by front surface double reflection as shown in Fig. 2.16. Geometrical light-trapping occurs for infrared (IR) light, increasing the light's path and thus absorption probability in the c-Si bulk [Bre03].

2.4. Contact formation

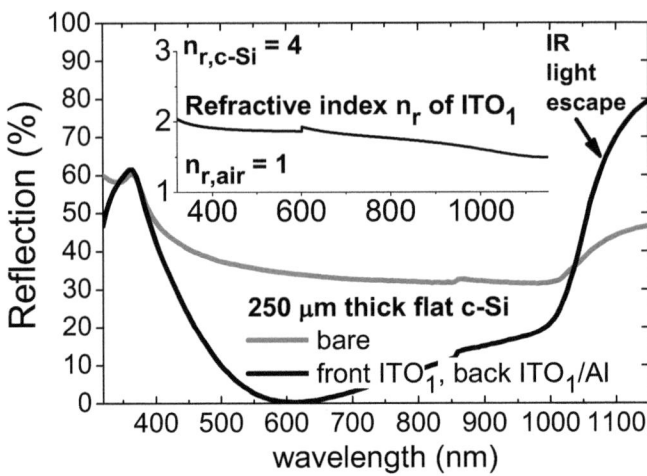

Figure 2.15: *Reflection reduction by 85 nm thick front ITO with wavelength dependent refractive index as shown in inset [DMP07] compared to the 250 μm thick bare flat c-Si surface's reflectivity.*

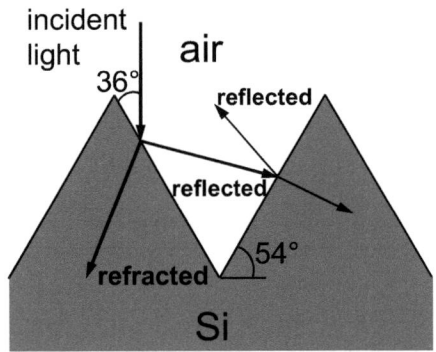

Figure 2.16: *Primary reflection reduction by double reflection (35% → 10% = (35%)²) on a textured Si surface where the surface grooves are formed by intersecting <111> equivalent planes forming pyramids.*

The reflectivity of bare Si is reduced from about 35% ($Refl = \frac{n_{c-Si}-n_{air}}{n_{c-Si}+n_{air}}$) for flat surfaces to around 10% ($\sim (35\%)^2$) for the textured surface as shown in Fig. 2.17 by the grey curves. Reflection, transmission and absorption of

2.4. Contact formation

a 180 µm thick wafer textured with large pyramids having a base length around 20 µm and a 220 µm thick wafer textured with small pyramids having a base length around 2 µm are identical. Compared to flat c-Si, the addition of an antireflection coating reduces the overall reflectivity of textured c-Si to a few percent on a much wider wavelength range, as shown in Fig. 2.17 by the black curves and thus increases J_{SC} in complete solar cells.

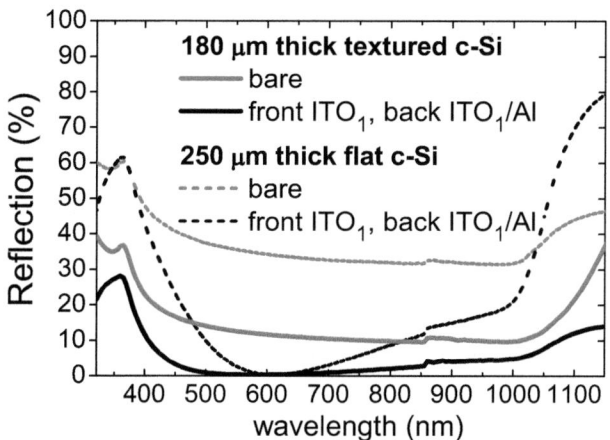

Figure 2.17: *Comparison of reflectance spectra from 250 µm thick flat (dashed lines) and 180 µm thick textured (straight lines) bare (gray lines) and 85 nm front ITO covered c-Si (black lines).*

Noticeably, an interesting candidate for the Si HJ solar cell application is hydrogen-doped In_2O_3. It has an excellent near-infrared-transparency due to its high free carrier mobility of up to 120 cm^2/(Vs) [KFK08]. But the indium based TCO's major drawback is the high cost and limited supply of indium. Possibly ITO could be replaced by ZnO which has, however, a lower conductivity [KCA$^+$07]. But today the first choice is still ITO because of its proven long term stability.

Structuring into individual solar cells Before ITO deposition on flat c-Si, individual solar cells of about (4.5 mm)^2 size are defined by masking the surface between them using a waterproof marker. After front and back surface contact deposition, the marker mask is dissolved in acetone including the ITO on top of it. This masking and lift-off leads to very well defined solar cells, which is all the more important for reaching a high FF, when the cells are small. Unfortunately, the marker stripes fizzle out on textured c-Si so that masking by marker is poor. Therefore, to define small individual solar cells on textured c-Si, a mask is mechanically fixed on top of the c-Si wafer during front ITO deposition.

2.4.2 Metallization

Our solar cells are very small and do not dispose of a front metal grid for the moment. On the back, aluminum (Al) or silver (Ag) were used as full-area back contacts. To prevent the detrimental metal diffusion into thin-film Si and to achieve good optical index-matching, an ITO layer was inserted between the BSF Si thin-film layer stack and the metal layer. ITO was deposited up to a thickness of 110 nm, but this thickness was never optimized. Al was DC-sputtered in the same deposition system as ITO (MRC system 603) while Ag was DC-sputtered in a Leybold Univex 450B system. Besides back contact formation, this TCO/metal layer stack serves as a back mirror increasing the lights' path length in the c-Si bulk, thus enhancing long-wavelength light's absorption probability. In Fig. 2.18, a textured c-Si wafer with a bare front shows that an ITO/Ag back more efficiently reflects infrared light, giving it a chance again to be absorbed, than an ITO/Al back reflector does. The best results would be achieved with a gold (Au) back reflector.

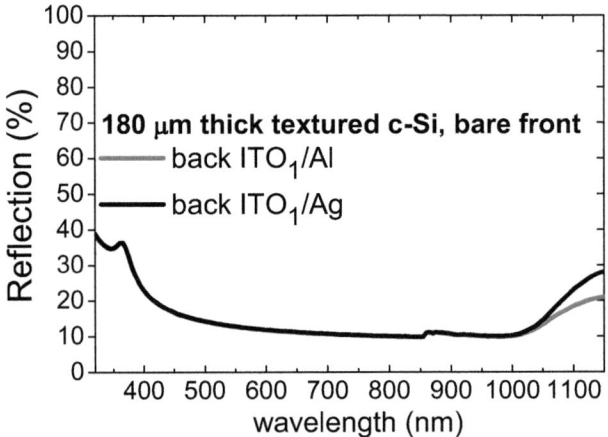

Figure 2.18: *Reflection spectra of 180 μm thick textured c-Si with an ITO/Al respectively an ITO/Ag back reflector and a bare front surface.*

2.5. Full a-Si:H/c-Si heterostructure processing

With the currently obtained IQEs at 1000 nm and considering that the c-Si's absorption coefficient starts to drop strongly from 1000 nm on (Fig. 2.2), the current gain to be expected from an ITO/Ag instead of an ITO/Al back reflector in our HJ solar cells is not significant. Therefore, because of the convenience of using the same deposition system (we simply place the substrate after ITO deposition in front of another target), an ITO/Al back contact is mainly used in our Si HJ solar cells.

2.5 Full a-Si:H/c-Si heterostructure processing

The maximal substrate size for the deposition system used is 8×8 cm^2. Therefore the round wafers with 100 mm diameter have to be cut. The minimal substrate size is given by the lifetime tester coil size that has a diameter of 3.5 cm. As there is no cleaning step except the HF-dip before the passivation layer deposition, wafers are touched as little as possible and cleaved by just scratching with the diamond stylus at the wafer edge. The native oxide removal is done using 4% HF diluted in DI water (Sec. 2.2.1), and stored at a volume of 3 l for multiple use. After the dip, a short DI-water shower rinse is carried out (if at all). Remaining droplets are blown away and the wafer piece is mechanically fixed on the substrate carrier. Usually, samples spend about 3 min in ambient air between coming out of the HF bath and going into the load-lock of the deposition system. This transfer time from the HF bath into vacuum is crucial because of fast reoxidation (Fig. 2.12(b)).

2.5.1 a-Si:H/c-Si passivation samples

For the fabrication of flat a-Si:H/c-Si passivation samples, double side polished float zone (FZ) 100 mm diameter wafers were purchased, mostly from the company Topsil [top]. By using such high-quality FZ wafers, we ensure a high bulk lifetime so that the effective carrier lifetime is limited by the a-Si:H/c-Si interface recombination (Eq. 2.5). For a-Si:H passivation such double side polished purchased c-Si wafers need no surface cleaning except a HF-dip to remove the native oxide forming on c-Si in ambient atmosphere (Sec. 2.2.1). The textured wafers need to be cleaned after texturing (Sec. 2.2.2), because this was not done by the texturers. After a wafer is put in vacuum as fast as possible to prevent reoxidation after the HF-dip, *i* a-Si:H layers are grown to standard thicknesses of around 45 nm to screen

2.5. Full a-Si:H/c-Si heterostructure processing

the influence of outer surface potential modification, i.e. intrinsic a-Si:H layer oxidation (Sec. 4.4.3). Stacked layer passivations are usually grown in the configurations such as the ones used for Si HJ solar cell emitter and BSF formation, i.e. 5 nm i a-Si:H + 15 nm doped layer (Sec. 2.3.4). Carrier lifetimes are almost exclusively characterized with the Sinton lifetime tester (Sec. 2.1.2.1), photogenerating carriers equally throughout the whole bulk by using an infrared filter. Neglecting bulk recombination, the effective surface recombination velocity S_{eff} is directly accessible from the measured effective lifetime τ_{eff}:

$$\frac{1}{\tau_{eff}} = \frac{1}{\tau_{surf}} = \frac{2S_{eff}}{W}, \qquad (2.22)$$

where W is the wafer thickness. As the effective surface recombination velocity S_{eff} [cm/s] is the sum of the front and back surface recombination velocities S_{front} and S_{back} [cm/s] ($S_{eff} = (S_{front} + S_{back})/2$), this implies for an unambiguous determination of the a-Si:H/c-Si interface recombination that both c-Si surfaces have to be identically passivated. Partially, in the aim of maximizing the symmetry of the deposited layers, a second HF-dip was carried out before growing the second's side passivation layer. However, mainly wafer pieces were just taken out of the load-lock, turned, fixed again and taken back under vacuum as fast as possible.

After having grown symmetrical thin-film Si passivation layers on both c-Si surfaces, the initial lifetime was measured by the Sinton lifetime tester. Finally, the passivated samples are annealed at 180 °C for 90 min under a nitrogen atmosphere (standard anneal for a-Si:H solar cells in our lab) and the lifetime is measured again. If not otherwise stated, lifetimes given are those of the annealed state. Such a low-temperature annealing is known to be beneficial for the interface passivation quality [DSH02, DWOB08].

2.5.2 a-Si:H/c-Si heterojunction solar cells

Up to the double-side thin-film Si growth the process sequence to fabricate a Si HJ solar cell is the same as for a passivation sample. Therefore, Fig. 2.19 summarizes the process described in Sec. 2.5.1, including the following contact deposition. Usually at least half of a 100 mm diameter wafer was used to keep a quarter in the state of a passivation sample without further contact deposition for lifetime measurements.

2.5. Full a-Si:H/c-Si heterostructure processing

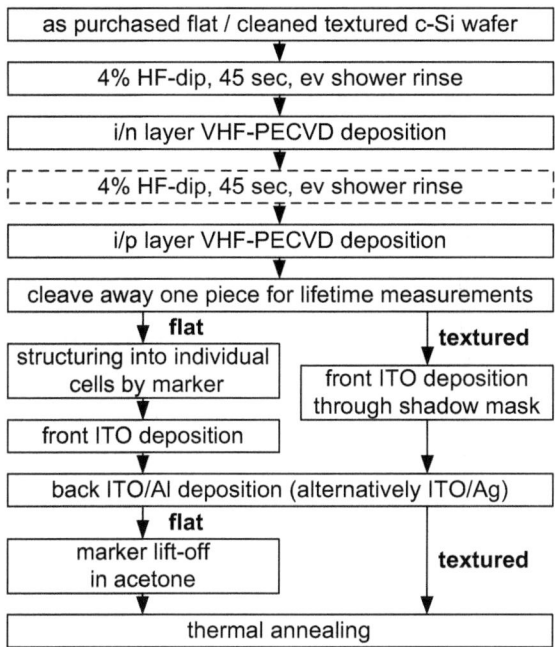

Figure 2.19: *Process flow for a-Si:H/c-Si heterojunction solar cell fabrication.*

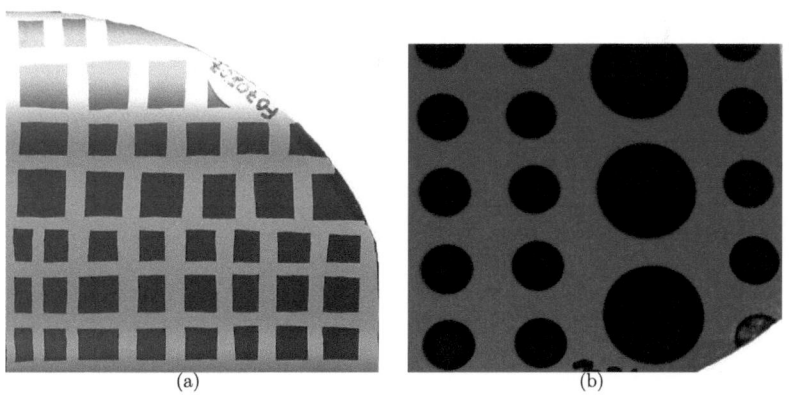

Figure 2.20: *Individual about (4.5 mm)² sized solar cells based on a) flat and b) textured c-Si including larger cells for comparison purposes.*

2.5. Full a-Si:H/c-Si heterostructure processing

For flat Si HJs the surface between individual solar cells is masked by marker before front ITO deposition, whereas for textured Si HJs a metal mask is mechanically fixed on the c-Si wafer for front ITO deposition. The wafer piece is turned and a full-area ITO/Al (or alternatively ITO/Ag) back contact is deposited. For flat cells, the marker mask is then dissolved in acetone. Figure 2.20 shows typical resulting wafer pieces with individual solar cells of about $(4.5\text{ mm})^2$ cell size, including larger cells for comparison purposes.

A standard anneal (180 °C, 90 min, N_2-atmosphere) is performed on the HJ solar cells as well as on the co-deposited passivation sample. On the former JV, EQE and SunsV_{OC} measurements are performed and on the latter, the lifetime is measured.

Chapter 3
Heterostructure interface recombination modeling

In the first part of this chapter (Sec. 3.2 to Sec. 3.4), the usual extended Shockley-Read-Hall (SRH) model is presented and discussed in detail for the case of SiO_2 and SiN_x passivation schemes on c-Si. The original recombination rate surface plot for this standard model is presented and the double-diode model is introduced. A detailed discussion of the effect of the choices of the values of bulk lifetime (Sec. 3.2) on our evaluation of S_{eff} is given in Sec. 3.4. Modeling of a-Si:H passivated c-Si interfaces is presented in Sec. 3.5 and Sec. 3.6.

3.1 Introduction

The historical development of the metal-oxide-semiconductor field-effect transistor (MOSFET) was only made possible thanks to the development of high-quality thick silicon dioxide (SiO_2) passivation layers [ATS59, Kah76]. Ever since the invention of the transistor [BB48], surface effects were extensively studied [Bro53]. Grove and Fitzgerald [GF66, FG68] extended the MOS theory to the case in which the surface space-charge region is not in thermal equilibrium. Using a quasi-exact solution of the surface recombination rate by means of the approximation of flat electron and hole quasi-Fermi levels in the space-charge region, this theory was later on improved by Girisch et al. [GMDK88] and named "extended SRH formalism".

The microscopic nature of the interface defects responsible for surface recombination has been identified as dangling bonds [BJS+85, KLK88], where recombination-specific properties are determined by the overlying passivating material.

3.1. Introduction

The purpose of interface recombination modeling is to get insight into a material's specific surface recombination properties by comparison to experimental values. The experimentally accessible value is the effective lifetime τ_{eff} (Sec. 2.1.2.1) that is given by bulk and front/back surface recombination:

$$\frac{1}{\tau_{eff}} = \frac{1}{\tau_{bulk}} + \frac{S_{front}}{W} + \frac{S_{back}}{W}. \qquad (3.1)$$

S_{front} [cm/s] and S_{back} [cm/s] denote the front and back surface recombination velocities respectively and W [cm] is the wafer thickness. The surface recombination velocity S [cm/s] is introduced to describe two-dimensional recombination and is related to the surface recombination rate U_s [cm^{-2}s^{-1}] by:

$$U_s \equiv S \times \Delta n_s, \qquad (3.2)$$

where Δn_s [cm^{-3}] is the excess carrier density at the interface. Equation 3.1 only holds when the bulk lifetime is sufficiently high enough to allow photogenerated carriers to reach both surfaces and when the surface recombination is sufficiently low enough to avoid transport-limited profiles near the surfaces [KS85, BRR03]. In this study, using high-lifetime FZ grown c-Si wafers and fabricating high-quality passivation layers, these two conditions are fulfilled throughout the experiments. Generally, these conditions can be assumed to occur if τ_{eff} is greater than 100 μs. In the case of symmetrical surface passivation, Eq. 3.1 reduces to two unknown parameters:

$$\frac{1}{\tau_{eff}} = \frac{1}{\tau_{bulk}} + 2 \times \frac{S}{W}, \qquad (3.3)$$

where $S_{front} = S_{back} = S$ and τ_{bulk} is given by:

$$\frac{1}{\tau_{bulk}} = \frac{1}{\tau_{extr}} + \frac{1}{\tau_{intr}} = \frac{1}{\tau_{defect}} + \frac{1}{\tau_{Aug}} + \frac{1}{\tau_{rad}}. \qquad (3.4)$$

Even if extrinsic bulk recombination τ_{extr} [s] is highly suppressed by using FZ wafers, i.e. the defect lifetime τ_{defect} [s] is very high, the bulk lifetime τ_{bulk} [s] is limited by intrinsic recombination processes (intrinsic lifetime τ_{intr} [s]). These are for an indirect-bandgap semiconductor Auger (Auger lifetime τ_{Aug} [s]) and radiative (radiative lifetime τ_{rad} [s]) recombination. Therefore, when measuring (high) τ_{eff}, we need to first calculate Auger and radiative recombination that may dominate interface recombination at high excess carrier densities to be able to extract S from Eq. 3.3.

The extended SRH formalism is used to model the two standard c-Si surface passivation schemes SiO$_2$ and silicon nitride (SiN$_x$). For more accurate modeling, it is necessary to determine the energy-dependance of the capture cross-sections and of the interface state density. This is usually done by deep level transient spectroscopy (DLTS) and capacitance-voltage (C-V) profiling [AGW92, SSS+97].

The extended SRH formalism is based on interface defects that have two possible charge states. Biegelsen *et al.* [BJS+85] identified common physical mechanisms underlying the characteristic attributes of the two systems a-Si:H and SiO$_2$/c-Si justified by their electron spin resonance (ESR) measurements. In bulk a-Si:H, recombination is based on amphoteric defects, i.e. dangling bonds having three charge states. In this study we extend a model previously established for bulk a-Si:H recombination [VJ86, HSS92] to the description of the surface recombination through dangling bonds. Within this model a few hypotheses can reduce the calculation of the recombination rate to the case of a discrete recombination level possessing three charge conditions where recombination occurs through four possible capture routes. The corresponding four different capture cross-sections are the microscopic recombination parameters governing recombination based on amphoteric defects. Therefore, we do not need to know the shape of the continuous distribution of dangling bond states resulting from the knowledge of the product of the capture cross-sections and the interface state density over the whole energy range in the bandgap [AGW92, SSS+97]. In summary, in this study we attempt to apply the most simple concepts to allow for a maximal modeling transparency.

3.2 Bulk recombination modeling: τ_{Aug}, τ_{rad} and τ_{defect}

Although high-lifetime wafers are used throughout this whole study, c-Si bulk recombination cannot be neglected due to Auger recombination dominating at high injection levels. For an accurate determination of S_{eff} from the measured τ_{eff}, an appropriate model for the injection level dependent c-Si bulk recombination has to be considered, like the one shown in this section. The concrete influence of the chosen c-Si bulk recombination model on S_{eff} is shown in Sec. 3.4.

Extrinsic bulk recombination processes are related to crystal defects. Intrinsic bulk recombination processes are inherent to the fundamental material properties and thus occur also in perfect semiconductor crystals. In the case of an indirect-bandgap semiconductor, such as c-Si, Auger and

3.2. Bulk recombination modeling: τ_{Aug}, τ_{rad} and τ_{defect}

radiative recombination are the two intrinsic bulk recombination processes. The lifetime limit imposed by c-Si bulk recombination τ_{bulk} [s] is therefore given by

$$\frac{1}{\tau_{bulk}} = \frac{1}{\tau_{extr}} + \frac{1}{\tau_{intr}} = \frac{1}{\tau_{extr}} + \frac{1}{\tau_{Aug}} + \frac{1}{\tau_{rad}}. \quad (3.5)$$

Auger recombination In the band-to-band Auger recombination process, an electron recombining with a hole transmits its excess energy to another electron or hole. Hence, as this recombination process involves three charge carriers, the higher the excess carrier densities and the wafer doping, the greater its efficiency. In high-quality c-Si with a good surface passivation, Auger recombination becomes the dominant recombination mechanism at high excess carrier densities. The most recent parametrization of the Auger recombination rate R_{Aug} [cm^{-3}s^{-1}] for varying doping and excess carrier densities has been proposed by Kerr and Cuevas by empirically fitting an extensive data set [KC]:

$$R_{Aug} = np(1.8 \times 10^{-24} n_0^{0.65} + 6 \times 10^{-25} p_0^{0.65} + 3 \times 10^{-27} \Delta n^{0.8}). \quad (3.6)$$

$n = n_0 + \Delta n$ [cm^{-3}] is the electron and $p = p_0 + \Delta p$ [cm^{-3}] the hole density, where n_0 and p_0 [cm^{-3}] are the electron and hole densities at thermal equilibrium conditions, and $\Delta n = \Delta p$ is the excess carrier density [cm^{-3}]. The product of the carrier densities at thermal equilibrium is given by the mass-action law $n_0 \times p_0 = n_i^2$, where n_i [cm^{-3}] is the intrinsic carrier density. The Auger lifetime τ_{Aug} [s] is then calculated using Eq. 3.6:

$$\tau_{Aug} = \frac{\Delta n}{R_{Aug}} = \frac{\Delta n}{np(1.8 \times 10^{-24} n_0^{0.65} + 6 \times 10^{-25} p_0^{0.65} + 3 \times 10^{-27} \Delta n^{0.8})}. \quad (3.7)$$

In Fig. 3.1(a), the excess carrier density dependent τ_{Aug} according to Eq. 3.7 is shown for different doping levels.

Radiative recombination Radiative recombination is the inverse process to optical generation, as an electron from the conduction band annihilates with a hole in the valence band, releasing its excess energy in the form of a photon with an energy close to the bandgap of the semiconductor. The efficiency of radiative recombination is thus, (as for Auger recombination,) directly proportional to the excess carrier densities and the wafer doping. The net radiative recombination rate R_{rad} [cm^{-3}s^{-1}] is given by [SMG74]

$$R_{rad} = 9.5 \times 10^{-15} np, \quad (3.8)$$

3.2. Bulk recombination modeling: τ_{Aug}, τ_{rad} and τ_{defect}

resulting in a radiative carrier lifetime τ_{rad} [s] of

$$\tau_{rad} = \frac{\Delta n}{R_{rad}} = \frac{\Delta n}{9.5 \times 10^{-15} np}. \tag{3.9}$$

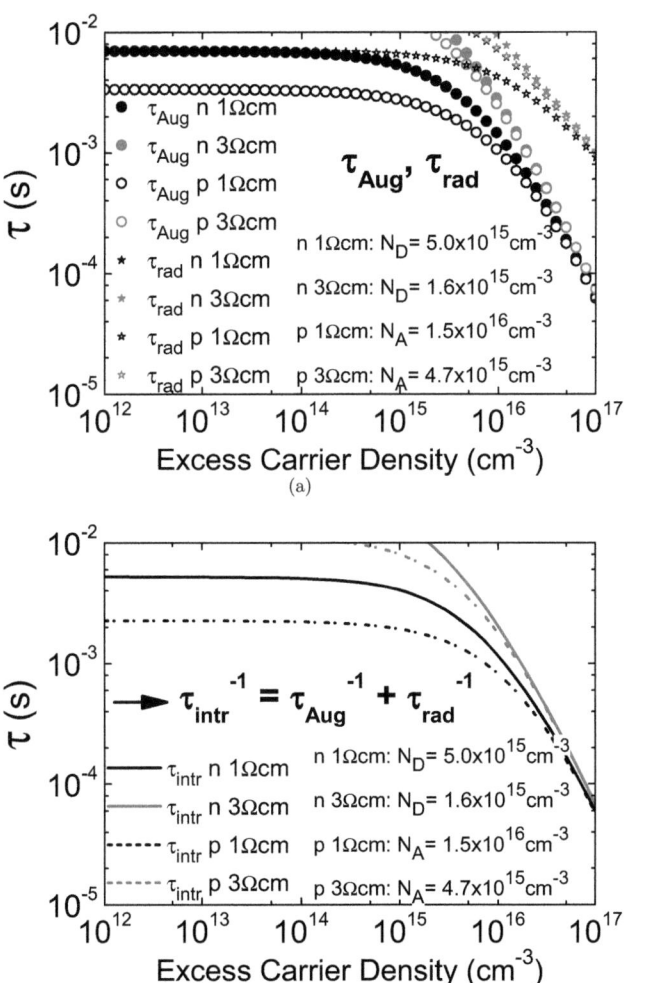

Figure 3.1: *a) Auger and radiative lifetime as a function of the excess carrier density for different doping levels which set b) the intrinsic carrier lifetime limit τ_{intr}.*

3.3. Standard interface recombination modeling

While radiative recombination is the dominant recombination process in direct-bandgap semiconductors, in an indirect-bandgap semiconductor, a phonon must be simultaneously emitted with a photon to conserve both energy and momentum. Such a four-particle process has a low probability to occur. This is the reason for the small value of the prefactor in Eq. 3.8. Figure 3.1(a) shows the excess carrier density dependance of τ_{rad} and Fig. 3.1(b) the intrinsic c-Si bulk lifetime τ_{intr} [s] resulting for different doping levels from

$$\frac{1}{\tau_{intr}} = \frac{1}{\tau_{Aug}} + \frac{1}{\tau_{rad}} = \frac{np}{\Delta n}(1.8 \times 10^{-24} n_0^{0.65} + 6 \times 10^{-25} p_0^{0.65} + 3 \times 10^{-27} \Delta n^{0.8} + 9.5 \times 10^{-15}). \quad (3.10)$$

Figure 3.1 shows that although Auger recombination dominates over radiative recombination, τ_{intr} in Eq. 3.10 is overestimated when neglecting radiative recombination.

Extrinsic bulk recombination If the wafer supplier specifies an ingot lifetime τ_{ingot} [s], we set it equal to τ_{bulk} at $\Delta n = 1 \times 10^{15}$ cm^{-3}. Thus, the extrinsic lifetime (assumed to be injection level independent, which is a large oversimplification) is given by $\tau_{extr} = [\frac{1}{\tau_{ingot}} - \frac{1}{\tau_{intr}}(\Delta n = 1 \times 10^{15}$ cm$^{-3})]^{-1}$. Otherwise, we adopt the maximum measured lifetime value of Yablonovitch et al. [YAC+86] for the extrinsic (τ_{extr} [s]), i.e. defect lifetime (τ_{defect} [s]), $\tau_{extr} = \tau_{defect} = 37$ ms (the highest ever measured lifetime is 130 ms [TBH+04]).

3.3 Standard interface recombination modeling

First the basics of surface passivation including the standard SRH recombination via surface states having two possible charge states are introduced (Sec. 3.3.1). The so-called "extended SRH formalism" is then used to calculate heterostructure interface recombination under surface band bending conditions (Sec. 3.3.2). Interface recombination reduction is achieved either by reducing the interface defect density or by greatly reducing the density of one surface carrier type by field-effect. If the amount of field-effect passivation is very high, the double-diode model applies for the calculation of interface recombination (Sec. 3.3.3).

3.3.1 Shockley-Read-Hall interface recombination

Recombination through defect levels in semiconductors is usually described by the Shockley-Read-Hall (SRH) theory [SR52, Hal52]. An electron makes

3.3. Standard interface recombination modeling

a transition from the conduction band into the defect level in the bandgap and from there into the valence band where it recombines with a hole. Figure 3.2 shows the four possible interactions of electrons and holes with a defect level representing a recombination center located at a trap energy level E_t [eV]. The illustration is for the case of a recombination center with a single energy level that is neutral when not occupied by an electron and negatively charged when occupied by an electron. The four possible transitions are:

a) electron capture by an empty recombination center,

b) electron emission from an occupied recombination center,

c) hole capture by an occupied recombination center,

d) hole emission from an empty recombination center, i.e. excitation of an electron from the valence band to the unoccupied recombination center.

The single trap level recombination rate R_{SRH} [cm^{-3}s^{-1}] given by SRH theory is [Sze85]

$$R_{SRH}(n,p) = \frac{np - n_i^2}{\frac{n+n_i e^{(E_t-E_i)/kT}}{\sigma_p} + \frac{p+n_i e^{-(E_t-E_i)/kT}}{\sigma_n}} v_{th} N_t, \qquad (3.11)$$

given that the excess carrier densities of electrons and holes are equal, i.e. $\Delta n = \Delta p$. σ_n and σ_p [cm^2] are the capture cross-sections of electrons and holes, v_{th} [cm/s] is the thermal velocity of the charge carriers ($\sim 2 \times 10^7$ cm/s in Si at room temperature) and N_t [cm^{-3}] the trap density, i.e. defect density.

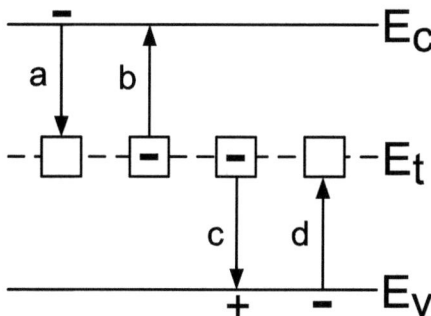

Figure 3.2: *Indirect recombination processes by trapping and thermal emission [SR52].*

3.3. Standard interface recombination modeling

When $n \times \sigma_n = p \times \sigma_p$, it follows from Eq. 3.11 that traps with energies E_t close to the center of the bandgap E_i [eV], are the most effective recombination centers. In analogy to bulk SRH recombination, the recombination rate U_{SRH} [cm^{-2}s^{-1}] at the c-Si surface via a single-level surface state having two possible charge states is given by

$$U_{SRH}(n_s, p_s) = \frac{n_s p_s - n_i^2}{\frac{n_s + n_i e^{(E_t - E_i)/kT}}{\sigma_p} + \frac{p_s + n_i e^{-(E_t - E_i)/kT}}{\sigma_n}} v_{th} D_{it}, \quad (3.12)$$

where n_s and p_s [cm^{-3}] are the electron and hole carrier densities at the surface (Eq. 3.20), and D_{it} [cm^{-2}] is the density of surface/interface traps. For a continuum of trap energy levels as found for example at SiO$_2$/c-Si interfaces [ES85], the energy-dependance of the surface state density D_{it} and of the capture cross-sections σ_n and σ_p have to be included, and the total recombination rate must be calculated via integration of the single-level recombination rates between the valence (E_V [eV]) and the conduction band edge (E_C [eV]):

$$U_{SRH}(n_s, p_s) = (n_s p_s - n_i^2) v_{th} \int_{E_V}^{E_C} \frac{D_{it}(E_t)}{\frac{n_s + n_i e^{(E_t - E_i)/kT}}{\sigma_p(E_t)} + \frac{p_s + n_i e^{-(E_t - E_i)/kT}}{\sigma_n(E_t)}} dE_t.$$
(3.13)

In analogy to the bulk carrier recombination lifetime $\tau \equiv \frac{\Delta n}{R}$ [s], a surface recombination velocity S [cm/s] is defined at the semiconductor surface as $U_s \equiv S \times \Delta n_s$ [cm^{-2}s^{-1}] where Δn_s [cm^{-3}] is the excess carrier density at the surface. Note that from a high recombination rate results a low lifetime τ, but a high surface recombination velocity S.

The excess carrier density dependent SRH surface recombination velocity S_{SRH} [cm/s] is given by Eqs. 3.12 and 3.2:

$$S_{SRH}(\Delta n_s) = \frac{1}{\Delta n} \frac{[(n_0 + \Delta n_s)(p_0 + \Delta p_s)] - n_i^2}{(n_0 + \Delta n_s) + n_i e^{(E_t - E_i)/kT} + \frac{\sigma_p}{\sigma_n}[(p_0 + \Delta p_s) + n_i e^{-(E_t - E_i)/kT}]} \sigma_p v_{th} D_{it},$$
(3.14)

provided that there are flatband conditions at the surface and therefore the excess surface carrier densities of electrons and holes are equal and equal also to the ones in the bulk, i.e. $\Delta n_s = \Delta p_s = \Delta n = \Delta p$. S_{SRH} depends, on one hand on the surface state properties D_{it}, E_t, σ_n and σ_p and, on the other hand, on the injection level Δn and on the wafer doping level n_0 (p_0). As follows from Eq. 3.14, D_{it} and σ_p only scale $S_{SRH}(\Delta n_s)$ by a constant factor.

3.3. Standard interface recombination modeling

The capture cross-section σ describes the effectiveness of a center to capture an electron (σ_n [cm^2]) or a hole (σ_p [cm^2]) and is thus a measure of how close the charge carrier has to come to the center to be captured. We expect that the capture cross-section would be in the order of the atomic dimensions, that is for silicon, in the order of 1×10^{-15} cm^2 [Sze85]. Therefore, we set $\sigma_p = 1 \times 10^{-16}$ cm^2 and $D_{it} = 1 \times 10^{10}$ cm^{-2} (typical values for silicon [Abe99]), together with the trap energy level E_t set at midgap. First we vary the doping level of a p-type c-Si wafer from an acceptor density N_A [cm^{-3}] $N_A = p_0 = 1 \times 10^{14}$ cm^{-3} to 1×10^{17} cm^{-3} while keeping the capture cross-section ratio $\frac{\sigma_n}{\sigma_p}$ constant. Figure 3.3(a) visualizes U_{SRH} from Eq. 3.12 as a function of n_s and p_s for equal capture cross-sections, i.e. $\frac{\sigma_n}{\sigma_p} = 1$. Note that in the case of flatband conditions, carrier densities are equal in the bulk and at the surface: $n_s = n = n_0 + \Delta n$ and $p_s = p = p_0 + \Delta n$. In the following experiments used for the characterization of surface passivation layers (described in Sec. 2.1.2.1), the ratio n_s/p_s varies. Concretely, in a Sinton lifetime measurement, at first, in high injection level conditions, $n_s \approx p_s \approx \Delta n$. As time proceeds, the illumination is either decreased exponentially (quasi-steady-state method) or it is switched off (transient method). As a consequence of recombination, n_s/p_s then varies with time and U varies accordingly, remaining over the surface plot of Fig. 3.3(a), e.g. in the trajectories. When varying Δn from 1×10^{17} cm^{-3} to 1×10^{12} cm^{-3}, such as typically is done in a Sinton lifetime measurement, n_s and p_s differ for different doping levels and the resulting trajectories on the surface recombination rate plot are thus not the same.

For 1×10^{17} cm^{-3} p-type doped c-Si, while Δn decreases, low injection level conditions ($\Delta n \ll p_0$) rapidly prevail and therefore $p_s \approx p_0$. It follows from Eq. 3.12 that $U_{SRH} \propto \Delta n$ and therefore $S_{SRH} = \frac{U_{SRH}}{\Delta n}$ becomes independent of the injection level as verified by the constant S_{SRH} value in the plot of $S_{SRH}(\Delta n)$ in Fig. 3.3(b). Contrariwise, for 1×10^{14} cm^{-3} p-type doped c-Si, the condition $\Delta n \gg p_0$ is easily fulfilled and thus high injection level conditions are quickly met, where $n_s = p_s = \Delta n$. The corresponding U_{SRH} is on the $n_s = p_s$–trajectory and it follows from Eq. 3.12 that $U_{SRH} \propto \Delta n$ again and therefore S_{SRH} becomes again independent of the injection level as shows Fig. 3.3(b).

3.3. Standard interface recombination modeling

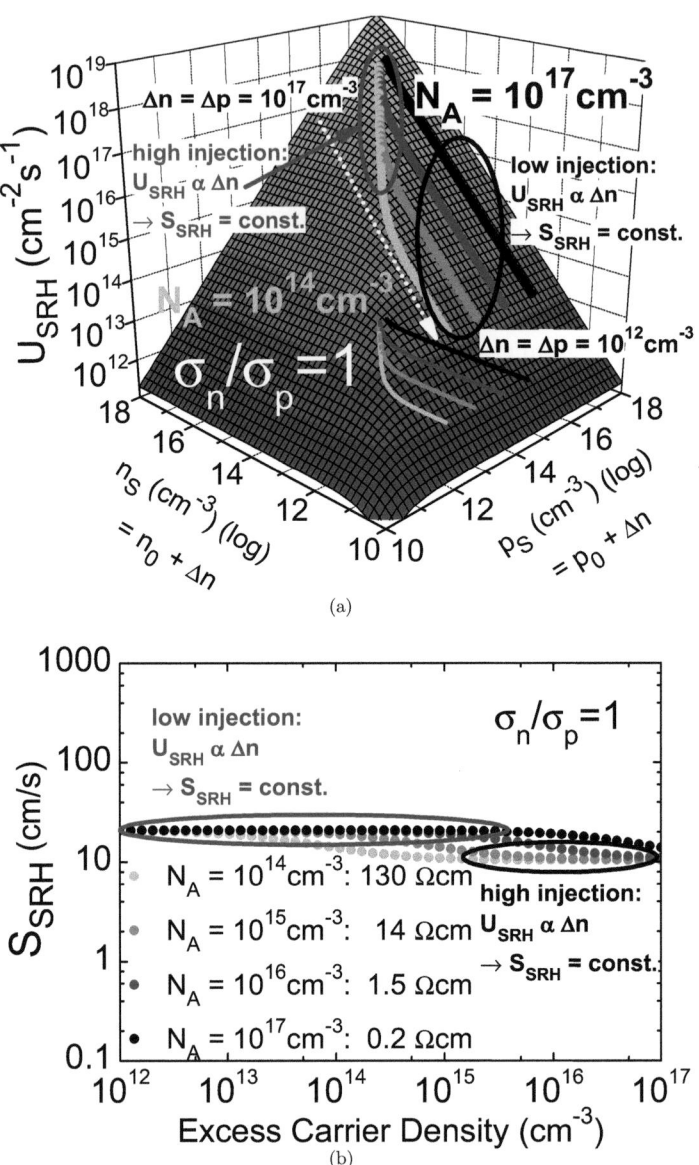

Figure 3.3: *Impact of the wafer doping level N_A on the injection level-dependance of a) the surface recombination rate and b) the surface recombination velocity for midgap surface defects with equal electron and hole capture cross-sections.*

3.3. Standard interface recombination modeling

A much higher influence on the injection level-dependance of U_{SRH} (and thus S_{SRH}) is given by varying the capture cross-section ratio $\frac{\sigma_n}{\sigma_p}$ while fixing the c-Si doping level, in this example to $p_0 = 1 \times 10^{16}$ cm^{-3}. U_{SRH} is maximal for $\frac{p_s}{n_s} = \frac{\sigma_n}{\sigma_p}$, as then the denominator in Eq. 3.12 becomes minimal. Figure 3.4 shows $U_{SRH} = \mathrm{f}(n_s, p_s)$-plots for $\frac{\sigma_n}{\sigma_p} = 1$, 50 and $\frac{1}{50}$. In the present case of flatband conditions, the variations of n_s and p_s with Δn are given by the doping level, $n_s = n_0 + \Delta n$ and $p_s = p_0 + \Delta n$, and therefore the n_s, p_s-projections in the surface plots of Fig. 3.4 are identical.

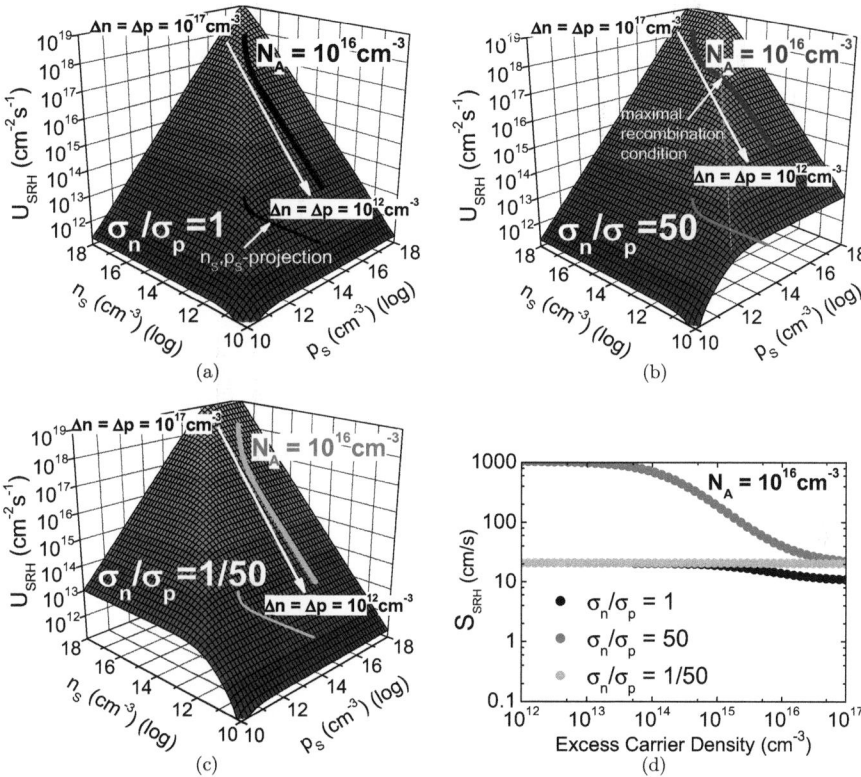

Figure 3.4: *Impact of the capture cross-section ratio $\frac{\sigma_n}{\sigma_p}$ on the injection level-dependance of the surface recombination rate on a p-type doped wafer ($p_0 = 1 \times 10^{16}$ cm^{-3}) with midgap-surface defects for a) $\frac{\sigma_n}{\sigma_p} = 1$, b) $\frac{\sigma_n}{\sigma_p} = 50$ and c) $\frac{\sigma_n}{\sigma_p} = \frac{1}{50}$. d) Resulting injection level dependent surface recombination velocity.*

But translating the common n_s, p_s-projection on the different U_{SRH} surface plots in Figs. 3.4(a), 3.4(b) and 3.4(c), $U_{SRH}(n_s, p_s)$ and thus $S_{SRH}(\Delta n)$ in Fig. 3.4(d) is significantly different. This is because $p_s > n_s$ for p-type doped c-Si at any injection level in flatband conditions. By consequence, U_{SRH} values at the maximal recombination condition $\frac{p_s}{n_s} = \frac{\sigma_n}{\sigma_p}$ can be reached only for $\frac{\sigma_n}{\sigma_p} = 50$ (Fig. 3.4(b)), as shown by comparison of the trajectories on the surface recombination rate plots in Figs. 3.4(a), 3.4(b) and 3.4(c).

3.3.2 Extended Shockley-Read-Hall interface recombination formalism

To reduce the recombination rate at the c-Si surface, there are two fundamentally different approaches which can be concluded from Eq. 3.12:

- Reduction of the interface defect density D_{it}: as from Eq. 3.12 $U_{SRH} \propto D_{it}$, reducing D_{it} decreases $S(\Delta n_s)$, over the whole excess carrier density range by the same factor.

- Reduction of the surface density of one carrier type: as the SRH recombination process involves one electron and one hole, recombination is highest when the surface electron and hole densities are about equal. As electrons and holes carry electrical charge, the formation of an internal electrical field below the semiconductor surface reduces the surface density of one charge carrier type. Technologically this is realized by either the implementation of a dopant profile underneath the c-Si surface or by field-effect passivation by means of stable electrostatic charges in an overlying insulator.

As a result of field-effect passivation, $\Delta n_s \neq \Delta p_s$. Thus, an effective surface recombination velocity S_{eff} [cm/s] is defined as

$$S_{eff} \equiv \frac{U_s}{\Delta n(x = d)}, \quad (3.15)$$

at a virtual surface within the wafer positioned at $x = d$, the edge of the c-Si surface space-charge region (Fig. 3.5), where the photogenerated bulk c-Si excess carrier densities are equal, i.e. $\Delta n(x = d) = \Delta p(x = d)$.

To examine the influence of the wafer doping level and the bulk excess carrier density ($\Delta n = \Delta p$) on S_{eff}, a simplified approach, the so-called

3.3. Standard interface recombination modeling

"extended SRH formalism" was introduced [GF66, FG68, GMDK88]. It permits to calculate the non-equilibrium surface carrier densities n_s and p_s from a given surface charge density Q_s [cm^{-2}] inducing a charge density Q_{Si} [cm^{-2}] in the c-Si surface (Fig. 3.5(a)). Recombination within this surface space-charge region is neglected.

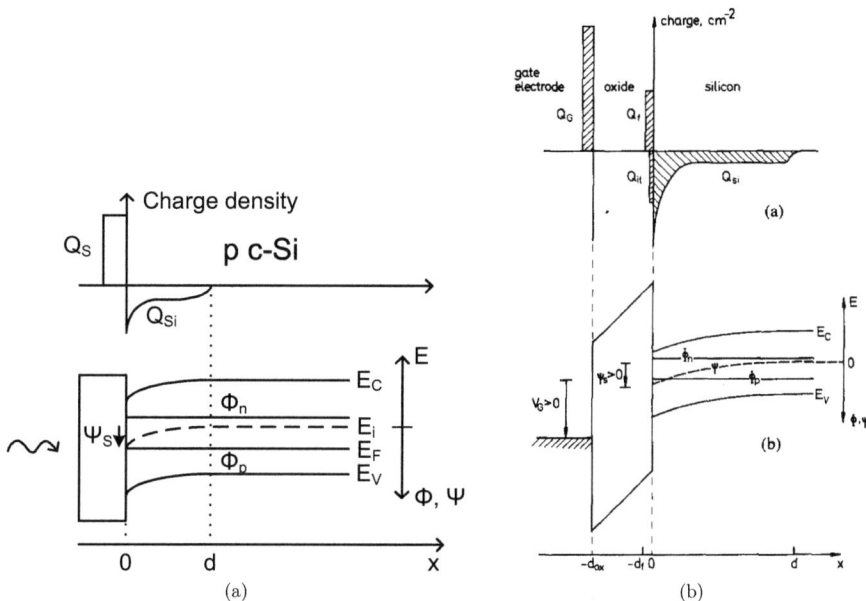

Figure 3.5: *Charge density and band diagram at the passivated c-Si interface under illumination, including definitions of electron energy and potentials. a) Charge density Q_{Si} induced in the c-Si surface by a surface charge density Q_s, composed for example in the case of a passivating oxide of b) the charge density in the interface traps Q_{it}, the fixed insulator charge density Q_f and the charge density induced in an optional gate electrode or deposited otherwise on the outer passivation layer's surface Q_G [GMDK88].*

More precisely, the total surface charge density Q_s is composed as shown in Fig. 3.5(b) of

$$Q_s = Q_{it} + Q_f + Q_G, \qquad (3.16)$$

where Q_{it} [cm^{-2}] is the charge density in the interface traps and Q_f [cm^{-2}] is the fixed insulator charge density. Q_G [cm^{-2}] is the charge density induced in the gate electrode or deposited otherwise on the outer passivation layer's

3.3. Standard interface recombination modeling

surface (for example by corona charging [SBLW94]). The charge in the interface traps can be written as [GMDK88]:

$$Q_{it} = -\int_{E_V}^{E_C} D_{it,A}(E) f_A(E) dE + \int_{E_V}^{E_C} D_{it,D}(E) f_D(E) dE, \quad (3.17)$$

where $D_{it,A}(E)$ and $D_{it,D}(E)$ [cm^{-2}eV^{-1}] are the acceptor and donor trap densities. The electron-occupation function of acceptor traps $f_A(E)$ [ ] and the hole-occupation function of donor traps $f_D(E)$ [ ] are given by:

$$f_A(E) = \frac{\sigma_n n_s + \sigma_p n_i e^{-(E_t - E_i)/kT}}{\sigma_n(n_s + n_i e^{(E_t - E_i)/kT}) + \sigma_p(p_s + n_i e^{-(E_t - E_i)/kT})},$$

$$f_D(E) = \frac{\sigma_p p_s + \sigma_n n_i e^{(E_t - E_i)/kT}}{\sigma_n(n_s + n_i e^{(E_t - E_i)/kT}) + \sigma_p(p_s + n_i e^{-(E_t - E_i)/kT})}. \quad (3.18)$$

The charge induced in the gate electrode is deduced from Faraday's law of induction:

$$Q_G(\psi_s) = -\frac{1}{d_{ins}}[Q_f \frac{d_f}{2} + \frac{\epsilon_0 \epsilon_{ins}}{q}(\psi_s - V_G)], \quad (3.19)$$

where V_G [V] is the bias applied to the gate electrode, d_f [cm] is the distance to the surface of the fixed charge's location, d_{ins} [cm] is the insulator thickness and ϵ_{ins} [ ] is the relative electrical permittivity of the insulator. The surface potential ψ_s [V] determines the interface carrier densities

$$n_s = (n_0 + \Delta n) e^{(+q\psi_s/kT)} \quad \text{and}$$
$$p_s = (p_0 + \Delta n) e^{(-q\psi_s/kT)}. \quad (3.20)$$

The charge density induced in the c-Si surface is given by

$$Q_{Si} = -\text{sign}(\psi_s) \sqrt{\frac{2kTn_i \epsilon_0 \epsilon_{Si}}{q^2}} [e^{q(\phi_p - \psi_s)/kT} - e^{q\phi_p/kT}$$
$$+ e^{q(\psi_s - \phi_n)/kT} - e^{-q\phi_n/kT} + \frac{q\psi_s(p_0 - n_0)}{kTn_i}], \quad (3.21)$$

where ϕ_n and ϕ_p [V] are the quasi-Fermi levels of electrons and holes at the edge of the space-charge region ($x = d$) that are given by

$$\phi_n = -(kT/q) \ln(\frac{n_0 + \Delta n}{n_i}) \quad \text{and}$$
$$\phi_p = +(kT/q) \ln(\frac{p_0 + \Delta n}{n_i}). \quad (3.22)$$

3.3. Standard interface recombination modeling

Q_{it} (Eq. 3.17), Q_G (Eq. 3.19) and Q_{Si} (Eq. 3.21) are all function of n_s and p_s, respectively ψ_s. The charge neutrality condition demands that:

$$Q_{Si} + Q_s = Q_{Si} + Q_{it} + Q_f + Q_G = 0. \quad (3.23)$$

ψ_s can be numerically approached by minimizing Eq. 3.23. We use MATLAB for a numerical approximation. The corresponding code is given in Appendix A.1. If the interface state density D_{it} is small, which is a necessary condition for good surface passivation, Q_{it} is negligible (Eq. 3.17), and if additionally, there is no charge density on the outer passivation layer's surface, i.e. $Q_G = 0$, it follows from Eq. 3.23 that $Q_s = Q_f = -Q_{Si}$. In this case, the non-linear equation

$$Q_s = \text{sign}(\psi_s)\sqrt{\frac{2kTn_i\epsilon_0\epsilon_{Si}}{q^2}[e^{q(\phi_p-\psi_s)/kT} - e^{q\phi_p/kT}} \\ \overline{+e^{q(\psi_s-\phi_n)/kT} - e^{-q\phi_n/kT} + \frac{q\psi_s(p_0-n_0)}{kTn_i}]} \quad (3.24)$$

can be numerically solved for ψ_s from Q_s for a given wafer doping and injection level (Eq. 3.22), and finally n_s and p_s are calculated with Eq. 3.20, i.e. $n_s, p_s = \text{f}(\Delta n; n_0, p_0; \psi_s(Q_s))$. Because the quasi-Fermi levels ϕ_n and ϕ_p depend on the non-equilibrium density of the respective carrier type (electron or hole) and are thus injection level dependent (Eq. 3.22), the surface potential ψ_s is also injection level dependent (Eq. 3.21).

More precisely, ψ_s decreases with increasing injection (i.e. illumination) level, and starts to vanish in high injection level conditions according to the magnitude of Q_s. This is because for $\Delta n > n_0, p_0$ the splitting of the quasi-Fermi levels is equally distributed around midgap, i.e. $|\phi_n| = |\phi_p|$ (Eq. 3.22). MATLAB is used to numerically solve the non-linear equation for ψ_s, i.e. Eq. 3.24, the corresponding code is given in Appendix A.2. Knowing the surface carrier densities n_s and p_s, the surface recombination rate U_{SRH} and the resulting effective surface recombination velocity $S_{eff,SRH}$ [cm/s] are given by Eqs. 3.12 and 3.15, i.e. $U_{SRH}, S_{eff,SRH} = \text{f}(\Delta n; n_0, p_0; Q_s, D_{it}; \sigma_n, \sigma_p)$.

Figure 3.6 shows the impact of a positive surface charge density on the trajectories on the surface recombination rate plot $U_{SRH} = \text{f}(n_s, p_s)$ together with the resulting $S_{eff,SRH}(\Delta n)$-curves. Again, p-type c-Si with $p_0 = 1 \times 10^{16}$ cm^{-3} and equal capture cross-sections are assumed. The previously discussed flatband condition case ($Q_s = 0$) serves as a starting point. Its n_s, p_s-projection is given in Fig. 3.6(a). The positive surface charge repels holes from the surface. As a result, for the smallest charge

59

3.3. Standard interface recombination modeling

($Q_s = 5 \times 10^{10}$ cm^{-2}), surface electrons become just a bit more numerous than holes. For the highest charge in our example ($Q_s = 50 \times 10^{10}$ cm^{-2}), the surface is in strong inversion, i.e. it has become n-type and due to the absence of holes, the resulting $S_{eff,SRH}$ in Fig. 3.6(b) is very low. For intermediate charge densities ($Q_s = 10 \times 10^{10}$ cm^{-2} for $p_0 = 1 \times 10^{16}$ cm^{-3}, discussion below), the hole density is slightly reduced and the electron density increased (depletion) such as $n_s \approx p_s$ and the maximal recombination condition $\frac{p_s}{n_s} = \frac{\sigma_n}{\sigma_p}$ is reached. However, at high injection levels, the $S_{eff,SRH}(\Delta n)$-curves in Fig. 3.6(b) approach the flatband curve as $\Delta n > p_0$, and therefore high injection level conditions prevail, where ψ_s starts to vanish according to the magnitude of Q_s.

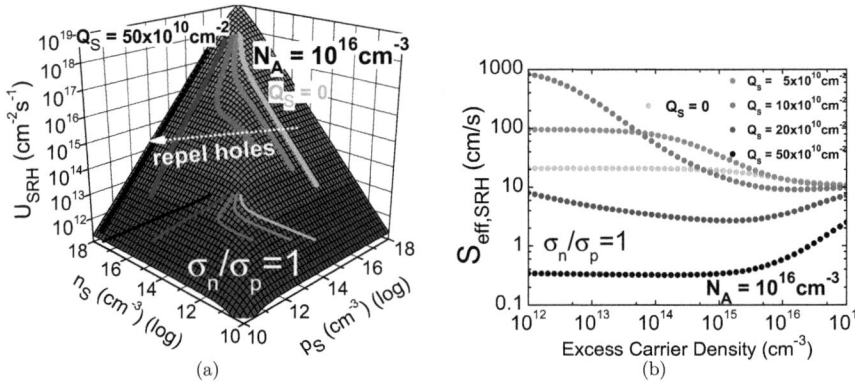

Figure 3.6: *Impact of a positive surface charge density on the injection level-dependance of a) the trajectories on the surface recombination rate plot and b) the resulting effective surface recombination velocity curves on a p-type doped wafer with midgap surface defects and equal electron and hole capture cross-sections.*

Finally, Fig. 3.7 shows again (as does Fig. 3.3) the recombination-dependance on the doping level, but with a positive surface charge of $Q_s = 10 \times 10^{10}$ cm^{-2} instead of flatband conditions, that shows a much more pronounced injection level-dependance (compare Figs. 3.7 and 3.3). Again, the n_s, p_s-projections of the trajectories on the surface plots in Fig. 3.7(a) (and therefore the $S_{eff,SRH}(\Delta n)$-curves) can be intuitively interpreted: for the highest doping density, the positive surface charge of $Q_s = 10 \times 10^{10}$ cm^{-2} is not sufficiently high to effectively repel the numerous holes from the surface, and $S_{eff,SRH}$ in Fig. 3.7(b) is thus unchanged

3.3. Standard interface recombination modeling

as compared to the flatband case in Fig. 3.3(b), while in the case of low doping, the given amount of positive surface charge very effectively repels the less numerous holes from the surface, such as to cause an inversion at the surface that greatly reduces $S_{eff,SRH}$ in Fig. 3.7(b).

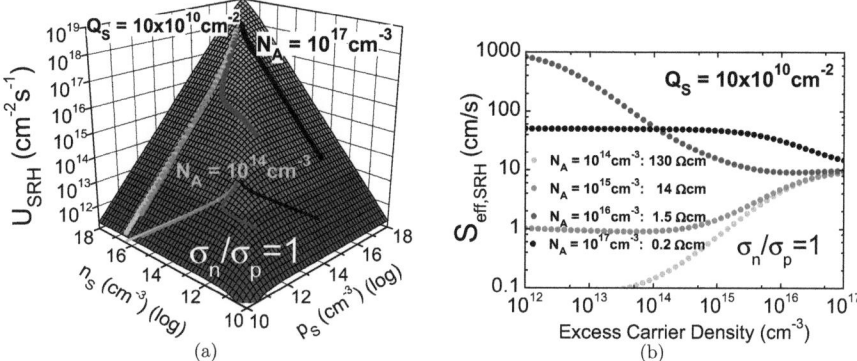

Figure 3.7: *Impact of the wafer doping level N_A for a given positive surface charge density on the injection level-dependance of a) the surface recombination rate and b) the effective surface recombination velocity for midgap surface defects with equal electron and hole capture cross-sections.*

Note that, while the doping level is easily known, strongly injection level dependent $S_{eff,SRH}(\Delta n)$-curves are caused i) either by asymmetrical capture cross-sections in the absence of surface charge (flatband conditions, Fig. 3.4(d)) or ii) by the presence of a surface charge (with symmetrical capture cross-sections, Fig. 3.6(b)).

The extended SRH formalism neglects recombination within the surface space-charge layer. If, during the passivation layer's growth, surface damage is produced, such as that observed, for instance, during the PECVD nitridation [SZR+05], this approximation is no longer valid and recombination occurring within the surface space-charge region becomes an important additional recombination path. Dauwe calculates recombination in the surface space-charge region using the theory of recombination via defect levels developed by Shockley, Read and Hall, by assigning the free charge carriers a much lower carrier lifetime than in the quasi-neutral bulk [Dau04]. Kerr models recombination occurring within an emitter region using a recombination current flowing into the emitter [Ker02] as presented in the next section (Sec. 3.3.3).

3.3. Standard interface recombination modeling

3.3.3 Emitter and BSF recombination: double-diode modeling

If the fundamental properties of the defects at the Si/insulator interface are known, it is possible to model the $S_{eff}(\Delta n)$-dependance using the extended SRH formalism. In its form developed in Sec. 3.3.2, the extended SRH formalism neglects recombination in the space-charge region. If the fundamental properties are not known, but Q_s is sufficiently high to create inversion conditions at the surface (Figs. 3.6 and 3.7 including related discussions), the surface behaves like an emitter region and the surface recombination current J_{rec} [mAcm^{-2}] can be written as [Ker02]:

$$J_{rec} = qWR_{rec} = 2qS_{eff,2diode}\Delta n = 2J_{01}(e^{\frac{qV}{n_{01}kT}} - 1) + 2J_{02}(e^{\frac{qV}{n_{02}kT}} - 1), \quad (3.25)$$

and therefore within this double-diode model

$$S_{eff,2diode}(\Delta n) = \frac{J_{01}}{q\Delta n}[(\frac{(n_0 + \Delta n)(p_0 + \Delta n)}{n_i^2})^{\frac{1}{n_{01}}} - 1] + \frac{J_{02}}{q\Delta n}[(\frac{(n_0 + \Delta n)(p_0 + \Delta n)}{n_i^2})^{\frac{1}{n_{02}}} - 1], \quad (3.26)$$

with the diode ideality factors $n_{02} > n_{01}$ [ ], where J_{01} and J_{02} [mAcm^{-2}] determine the recombination currents associated with n_{01} and n_{02}. In its simplest interpretation, J_{01} with $n_{01} = 1$ is assigned to the surface passivation quality, while J_{02} with $n_{02} = 2$ is assigned to recombination in the space-charge region [Sze85]. A higher J_{01} leads to lower V_{OC}s, while a higher J_{02} will lower voltages around the maximum power point, thus lowering FF. For illustration purposes we consider a positive surface charge, sufficiently high to yield inversion conditions on a 1×10^{15} cm^{-3} doped p-type wafer (15 Ωcm). Figure 3.8(a) shows $S_{eff,2diode}(\Delta n)$-curves calculated from Eq. 3.26 with $J_{01} = 30$ pA/cm^2 ($n_{01} = 1$) and $J_{02} = 5$ nA/cm^2 ($n_{02} = 2$) on this p-type wafer as well as on an equivalently n-type doped wafer ($N_D = 1 \times 10^{15}$ cm^{-3} ~ 5 Ωcm). For BSF passivation, i.e. accumulation conditions (at the surface the majority charge carrier type is much more numerous than the minority one) the maximum recombination condition $\frac{p_s}{n_s} = \frac{\sigma_n}{\sigma_p}$ is never reached (Sec. 3.3.2) and the second term in Eq. 3.26 (accounting for recombination in the space-charge region) is very small. For emitter passivation, recombination within the space-charge region causes the $S_{eff,2diode}(\Delta n)$-curve to be shifted to higher values at low injection levels.

3.3. Standard interface recombination modeling

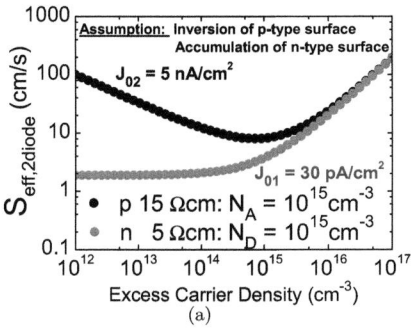
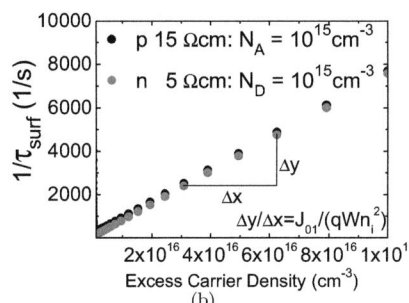

Figure 3.8: *a)* $S_{eff,2diode}(\Delta n)$*-dependance as calculated from Eq. 3.26 for p- and n-type silicon, assuming a positive charge density sufficiently high to result in inversion of the p-type surface, respectively accumulation of the n-type surface. b) Double linear scale* $\frac{1}{\tau_{surf}}(\Delta n)$*-dependance in high injection conditions such as plotted from lifetime measurements to extract* J_{01} *from Eq. 3.27.*

Provided the wafer is in high injection ($\Delta n > p_0, n_0$), J_{01} can be accurately measured from the slope of the $\frac{1}{\tau_{surf}} = f(\Delta n)$-plot shown in Fig. 3.8(b), as given in [KS85]

$$\frac{1}{\tau_{surf}} = \frac{J_{01}}{qWn_i^2}\Delta n. \qquad (3.27)$$

One has to keep in mind that many real effects such as injection level dependent surface recombination are not taken into account in this double-diode model. Therefore, the J_{01} and J_{02} values determined by fitting to experimental $S_{eff}(\Delta n)$-curves depend on the injection level-dependance of the surface recombination.

3.4. Determination of interface recombination parameters: interface recombination center density, field-effect passivation

3.4 Determination of interface recombination parameters: interface recombination center density, field-effect passivation

For comparison to modeled interface recombination, the measured S_{eff} ($S_{eff,m}$) is determined from the experimentally accessible τ_{eff} ($\tau_{eff,m}$). The choice of the c-Si bulk recombination parametrization (Sec. 3.2) is decisive for the value of $S_{eff,m}$ at high injection levels as will be shown in the first part of this section. As a consequence of this incertitude in the determination of $S_{eff,m}$ from $\tau_{eff,m}$, measured and calculated $\tau_{eff}(\Delta n)$-curves are compared further on, except where the absolute value of $S_{eff,m}$ is of major interest. The extended SRH formalism is applied to model SiO_2/c-Si interface recombination (3.4.2.1), while the charge measured in SiN_x is so high that the double-diode model is valid for interface recombination modeling on almost all c-Si doping levels (3.4.2.2).

3.4.1 Issues when comparing modeled and measured injection level dependent recombination curves

The parameters governing interface recombination such as described by the extended SRH model are:

- the capture cross-sections σ_n and σ_p, moreover their ratio $\frac{\sigma_n}{\sigma_p}$, describing the effectiveness of a recombination center to capture an electron or a hole,

- the total density of interface recombination centers N_s [cm^{-2}], synonym for the quality of the passivation,

- and the surface charge density Q_s, representative of the magnitude of the field-effect passivation.

The energy-dependance of the capture cross-sections $\sigma_{n,p}(E)$ and of the interface state density $D_{it}(E)$ can be determined by deep level transient spectroscopy (DLTS) [Lan74]. The fixed charge density Q_s of dielectric films is usually determined by capacitance-voltage (C-V) measurements [HG60]. For both, a metal-insulator-semiconductor (MIS) structure is formed and the surface potential ψ_s is changed by varying the gate voltage. While

3.4. Determination of interface recombination parameters: interface recombination center density, field-effect passivation

this requires the additional deposition of a metallic gate contact, the surface potential can also be changed by the deposition of additional charges upon the insulating film, such as corona charges [SBLW94]. A controlled charging permits then to evaluate the fixed charge density, because when the deposited corona charge density annihilates the oxide charge density, recombination is maximal (maximal SRH recombination rate for $\frac{p_s}{n_s} = \frac{\sigma_n}{\sigma_p}$ (Sec. 3.3.2)).

An alternative to these experimental techniques is to estimate $\frac{\sigma_n}{\sigma_p}$, N_s and Q_s by measuring $S_{eff,m}(\Delta n)$-curves of the same surface passivation layer on differently doped c-Si wafers and fitting the set of experimental $S_{eff,m}(\Delta n)$-curves with the by the extended SRH formalism calculated $S_{eff,SRH,c}(\Delta n)$-curves.

If, however the c-Si surface is in inversion or alternatively in accumulation over the whole injection level range, because of an emitter or BSF formation or because of a sufficiently large surface charge, $S_{eff}(\Delta n)$-curves can be modeled by the double-diode model. In this case the injection level-dependance of surface recombination is neglected.

The reason for generally representing $S_{eff}(\Delta n)$-curves rather than $\tau_{eff}(\Delta n)$-curves is the need for a direct comparison of only interface recombination, on all sorts of c-Si wafers, independently from the wafer thicknesses and differing bulk recombination. Neglecting bulk recombination in first approximation, the wafer thickness directly scales $S_{eff,m}$, i.e. doubling the wafer thickness doubles $S_{eff,m}$ when the same $\tau_{eff,m}$ is measured (Eq. 3.3). But even on wafers with the same thicknesses (300 μm is assumed here), a measured $\tau_{eff,m}$ of 1 ms can for example correspond to $S_{eff,m}$ differing by a factor 2, because intrinsic c-Si bulk recombination (Auger and radiative) depends on the wafer doping and on the injection level that this lifetime is measured (Eq. 3.3 and Sec. 3.2), as is shown by the following four examples:

- 1 Ωcm n-type, $\Delta n = 1 \times 10^{15}$ cm^{-3}, $\tau_{eff,m} = 1$ ms → $S_{eff,m} = 11$ cm/s
- 1 Ωcm n-type, $\Delta n = 5 \times 10^{15}$ cm^{-3}, $\tau_{eff,m} = 1$ ms → $S_{eff,m} = 7$ cm/s
- 10 Ωcm n-type, $\Delta n = 1 \times 10^{15}$ cm^{-3}, $\tau_{eff,m} = 1$ ms → $S_{eff,m} = 14$ cm/s
- 10 Ωcm n-type, $\Delta n = 5 \times 10^{15}$ cm^{-3}, $\tau_{eff,m} = 1$ ms → $S_{eff,m} = 12.5$ cm/s

However, when very high lifetimes are measured, the determination of the accurate contribution by surface recombination to the measured lifetime is rather difficult in the high excess carrier density range where intrinsic c-Si recombination becomes the dominant recombination mechanism. For illustration purposes, Fig. 3.9 shows the $S_{eff,m}(\Delta n)$-curve of our highest measured lifetime (n-type c-Si, 28 Ωcm), calculated from the $\tau_{eff,m}(\Delta n)$-curve with Eq. 3.3, assuming:

3.4. Determination of interface recombination parameters: interface recombination center density, field-effect passivation

a) by dots τ_{bulk} given by Kerr's intrinsic c-Si recombination parametrization, (τ_{intr} from Eq. 3.10 (Sec. 3.2)), and Yablonovitch's extrinsic upper lifetime limit, $\tau_{defect} = 37$ ms [YAC+86],

b) only Yablonovitch's minimum extrinsic recombination, resulting in $\tau_{bulk} = \tau_{defect} = 37$ ms, neglecting all intrinsic c-Si recombination,

c) and finally no c-Si bulk recombination at all, i.e. $\tau_{bulk} = \infty$.

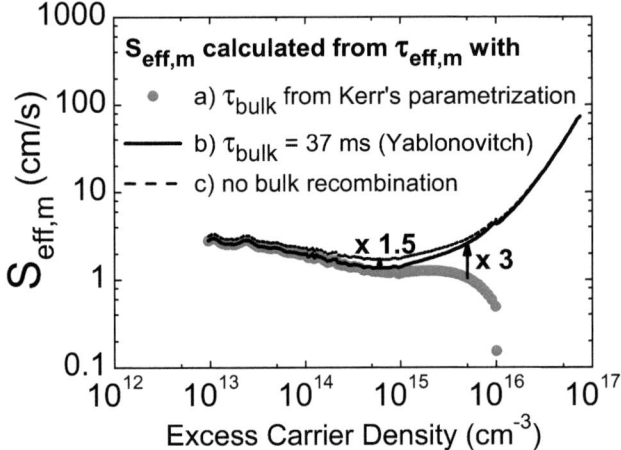

Figure 3.9: $S_{eff,m}(\Delta n)$-curve calculated with Eq. 3.3 from the $\tau_{eff,m}(\Delta n)$-curve measured on a 28 Ωcm n-type doped wafer, assuming a) by grey dots intrinsic and extrinsic c-Si bulk recombination (Kerr's parametrization, Sec. 3.2), b) by the full black line only extrinsic bulk recombination and c) by the dashed black line no c-Si bulk recombination at all. We observe a high sensitivity to the assumed intrinsic bulk lifetime parametrization at high excess carrier densities and differences in the minimal $S_{eff,m}$ value due to the assumption of the extrinsic lifetime value.

When neglecting intrinsic c-Si bulk recombination, at $\Delta n = 5 \times 10^{15}$ cm^{-3} interface recombination is overestimated by almost a factor of 3 (compare in Fig. 3.9 the symbol curve with the line curves). For $\Delta n > 1 \times 10^{16}$ cm^{-3}, $S_{eff,m}$ as calculated including intrinsic c-Si recombination starts to become negative (dot symbol curve in Fig. 3.9), but $\tau_{eff} > \tau_{bulk}$ is physically not possible. There are several issues that can cause $\tau_{eff,m}$ to exceed the calculated τ_{bulk}, among them Kerr's assumption used for parameterizing

3.4. Determination of interface recombination parameters: interface recombination center density, field-effect passivation

τ_{bulk} that Yablonovitch's and his own passivation [KC] is perfect at very high carrier densities. Thus, our data actually confirm Kerr's high injection lifetime limit quite closely, within uncertainties (Fig. 3.10).

In addition, when no bulk lifetime is specified by the wafer supplier, the calculation of the minimal $S_{eff,m}$ from the maximal $\tau_{eff,m}$ is very sensitive to the assumed maximal lifetime. Already with the very high assumed extrinsic c-Si lifetime of $\tau_{defect} = 37$ ms, the minimum $S_{eff,m}$ decreases by a factor of 1.5 as compared to the assumption of an infinite c-Si bulk lifetime (compare in Fig. 3.9 the full and the dashed line curves at their minima).

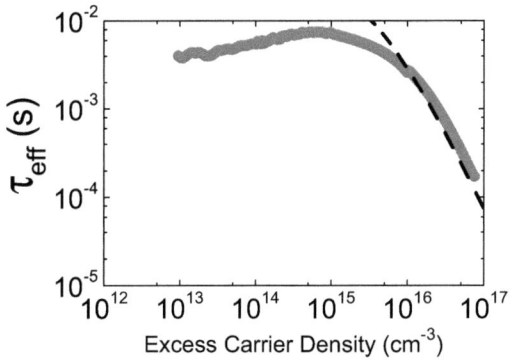

Figure 3.10: *On 28 Ωcm n-type doped c-Si measured $\tau_{eff,m}(\Delta n)$-curve (grey dots) and τ_{bulk} (dashed line) calculated as described in Sec. 3.2: Auger recombination dominates over surface recombination at high excess carrier densities and hinders the accurate determination of the contribution of surface recombination to the total recombination. Thus, when measuring very high lifetimes, $\tau_{eff}(\Delta n)$- including τ_{bulk}-curves will be plotted further on instead of $S_{eff}(\Delta n)$-curves.*

Because of these incertitudes in the accurate determination of the contribution of surface recombination to the total recombination, we will plot $\tau_{eff,m}(\Delta n)$-curves including τ_{bulk} from Sec. 3.2 when measuring very high lifetimes instead of $S_{eff,m}(\Delta n)$-curves, as shown in Fig. 3.10.

For the simultaneous representation of the calculated interface recombination $S_{eff,c}$, Eq. 3.3 is used together with the bulk lifetime parametrization from Sec. 3.2, to calculate the calculated effective lifetime $\tau_{eff,c}$:

$$\tau_{eff,c} = (\frac{1}{\tau_{bulk}} + 2 \times \frac{S_{eff,c}}{W})^{-1}. \tag{3.28}$$

3.4. Determination of interface recombination parameters: interface recombination center density, field-effect passivation

3.4.2 Modeling the standard c-Si surface passivation schemes SiO_2 and SiN_x

In this section, the extended SRH recombination formalism (Sec. 3.3.2) and the double-diode model (Sec. 3.3.3) are applied to the SiO_2/c-Si (Sec. 3.4.2.1) respectively to the SiN_x/c-Si interface (Sec. 3.4.2.2). Within the current interpretation, the injection level-dependance of interface recombination at the SiO_2/c-Si interface is given by asymmetrical capture cross-sections and a slightly positive surface charge density [GSWW], while recombination at the SiN_x/c-Si interface is supposed to be dominated by the high positive surface charge density in combination with an enhanced recombination in the induced surface space-charge region [Ker02, Dau04]. The resulting injection level-dependance of interface recombination is very similar (SiO_2 in Fig. 3.11 and SiN in Fig. 3.13). In Sec. 3.6 we will show that the passivation mechanisms of Si based layers on c-Si can alternatively be interpreted by considering the amphoteric nature of Si interface dangling bonds, still in combination with a surface charge density (Sec. 3.5).

3.4.2.1 Example SiO_2

In 1974 Green's [GKS74] metal-insulator-semiconductor (MIS) solar cells were the first solar cells benefitting from the passivating effect of silicon heterostructures, in this case from silicon dioxide (SiO_2). The use of SiO_2 for passivating very lightly diffused surfaces resulted in the first Si solar cells with efficiencies over 21% [KSS89]. Today, state of the art surface passivation for laboratory record efficiency c-Si solar cells still uses SiO_2 fabricated by thermal oxidation at temperatures around 1000 °C. The interface state density of as-grown SiO_2 passivation layers can be reduced by a low-temperature (400 °C) forming gas anneal (FGA: N_2 with a few percent of H_2). At such FGA annealed SiO_2/c-Si interfaces the P_{b1} state [LMFW01] that is an oxide back bonded Si dangling bond, i.e. $Si_2O \equiv Si\cdot$ is the prevailing defect. Such extrinsic dangling bond defects with oxygen back bonds result in donor-like states within the Si bandgap. Donor-like states are unoccupied in equilibrium and thus positively charged. These unoccupied donor-states can be related to the fixed positive oxide charge Q_f at the SiO_2/c-Si interface. States at the SiO_2/c-Si interface are much more effectively reduced with the so-called "alneal" that consists of evaporating an Al layer onto the oxide followed by an anneal in N_2 at around 400 °C. By means of a corrosive reaction between residual water molecules within the oxide and the Al, the P_{b1} state density is strongly reduced. The re-

3.4. Determination of interface recombination parameters: interface recombination center density, field-effect passivation

sulting dominating defect is the intrinsic P_{b0} state that consists of Si back bonded dangling bonds Si ≡ Si· close to midgap. Additional defect are stretched Si−Si bonds, that lead to bandtail states forming only shallow defects in the Si bandgap and hence contributing only weakly to the surface recombination rate at SiO_2/c-Si interfaces. Therefore, alnealed SiO_2 mainly passivates silicon by a strong reduction of the interface recombination center density.

The determination of the capture cross-sections $\sigma_n(E)$ and $\sigma_p(E)$ of the P_{b0} state is subject to errors of one to two orders of magnitude due to the uncertainties with regard to the determination of the surface band bending [Abe99]. For the following example, an energy independent capture cross-section ratio of $\frac{\sigma_n}{\sigma_p} = 100$ (as is generally found in the literature [YSEW86, GBRW99]) is adopted to calculate the injection level dependent lifetime of Kerr's [Ker02] alnealed SiO_2 passivated wafer set of varying doping types and levels, shown in Fig. 3.11.

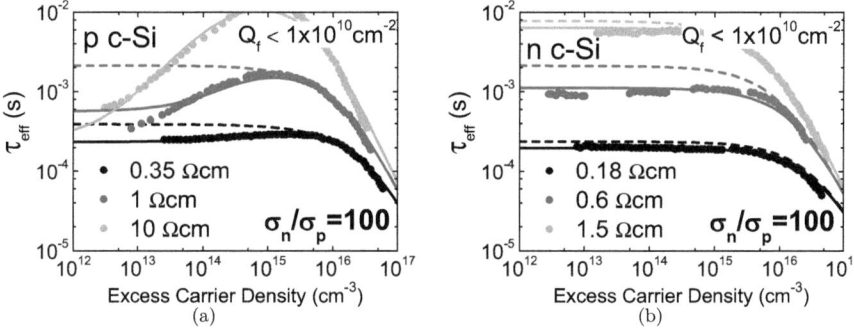

Figure 3.11: *Dots show measured effective lifetimes of Kerr's oxide-passivated, alnealed a) p- and b) n-type wafers of different doping levels [Ker02]. The upper lifetime limit as predicted by Kerr's parametrization is indicated by the dashed lines [KC]. Full lines show our fits to the experimental data with the extended SRH formalism (Sec. 3.3.2) assuming an energy independent capture cross-section ratio of $\frac{\sigma_n}{\sigma_p} = 100$, midgap surface-defects, negligible surface charge density and Kerr's bulk lifetime parametrization (Sec. 3.2).*

Q_f is attributed to extrinsic dangling bonds, the P_{b1} states, that form a sheet of positive charges very close to the SiO_2/c-Si interface. It is thus insensitive to the position of the Fermi level at the SiO_2/c-Si interface, i.e.

3.4. Determination of interface recombination parameters: interface recombination center density, field-effect passivation

Q_f can neither be altered by illumination nor by the doping level of the c-Si wafer base. Q_{it} is the charge density associated with the intrinsic dangling bond interface defects, the P_{b0} state, and depends thus on the Fermi level position. But at well-prepared SiO_2/c-Si interfaces, Q_{it} is much smaller than Q_f and, hence, can be neglected. Kerr's data is reasonably modeled by setting $Q_s = Q_f \leq 1 \times 10^{10}$ cm^{-2} that is smaller than the value range given by Aberle [Abe99] ($Q_f = 5 \times 10^{10} - 2 \times 10^{11}$ cm^{-2}). The injection level-dependance is given thus by the capture cross-section ratio $\frac{\sigma_n}{\sigma_p} = 100$, as shown in the surface plot in Fig. 3.12. Due to its asymmetrical capture cross-sections, SiO_2 yields a poorer passivation at low injection levels on p-type doped c-Si. In a Si solar cell, a SiO_2-passivation on p-type as compared to n-type c-Si yields the same high V_{OC} values but lower FF and J_{SC} values, as the Δn value at V_{OC} conditions is about 2 orders of magnitude higher than the Δn value at J_{SC} conditions.

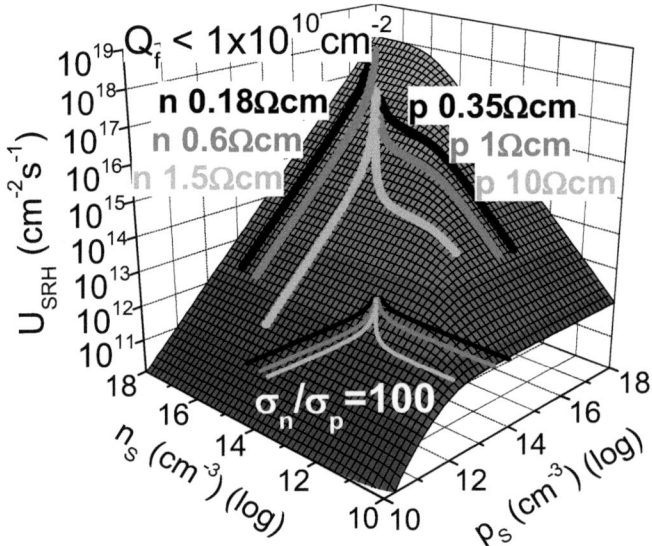

Figure 3.12: *Trajectories on the surface recombination rate plot illustrating the effect of asymmetrical capture cross-sections as the source of the measured injection level-dependance of the lifetimes in Fig. 3.11.*

The thermal oxidation of the c-Si surface is assumed to leave an undamaged SiO_2/c-Si surface, as the potentially damaged c-Si surface is partially consumed by the surface oxidation process. Therefore, neglecting recombi-

3.4. Determination of interface recombination parameters: interface recombination center density, field-effect passivation

nation in the surface space-charge layer is justified [GMDK88]. However, Sec. 3.4.2.2 will illustrate that very similar injection level dependencies of measured lifetimes are observed for SiN$_x$-passivated c-Si, which can also be very well modeled by the double-diode recombination model, i.e. including recombination in the space-charge region and neglecting the injection level-dependance of surface recombination.

3.4.2.2 Example SiN$_x$

Silicon nitride (SiN$_x$) passivation layers were first used the same as SiO$_2$ for metal-insulator-semiconductor devices [HSM80, HS81]. While the high temperatures needed to produce thermal SiO$_2$ passivation layers are unwanted in industrial processes for multiple reasons, PECVD deposited SiN$_x$ layers provide an outstanding surface passivation at deposition temperatures around 400 °C. The $\tau_{eff,m}(\Delta n)$-dependance at SiN$_x$/c-Si interfaces is astonishingly similar to the dependance measured at thermally grown SiO$_2$/c-Si interfaces. This is surprising, because the strong injection level-dependance of $S_{eff,m}$ would not have been expected from the large positive insulator charge $Q_f \sim 1 \times 10^{12}$ cm^{-2} measured in dark C-V and corona charging measurements of annealed samples. Since the conductivity of the SiN$_x$ films increases strongly with the Si-content, C-V measurements can no longer be conducted to determine Q_f in Si-rich surface passivation layers. Dauwe [Dau04] determined from sheet resistance data that the positive fixed interface charge decreases with an increasing Si-content of the films. Due to an interface state density reduction, Si-rich SiN$_x$ films result in higher effective carrier lifetimes than N-rich SiN$_x$ films. In contrast to other labs, Kerr and Cuevas [KC02a] succeeded in producing well-passivating stoichiometric Si$_3$N$_4$ films ($x = 1.33$), i.e. films with N-contents that would lead to poor lifetimes in other labs.

Little is know about the electronic properties of the involved defects at the SiN$_x$/c-Si interface, particularly the capture cross-sections. But from the observation of Sopori *et al.* that PECVD nitridation damages the c-Si surface [SZR$^+$05], it follows that recombination in the induced space-charge region becomes an important supplementary recombination path. For the case in which Q_f is sufficiently high, the band-bending results in the surface behaving like an induced emitter or a BSF region. Measured $\tau_{eff,m}(\Delta n)$-curves can thus be modeled with the double-diode recombination model (Sec. 3.3.3) as shown in Fig. 3.13. The model corresponds very well to the experiment except for the low injection range of the two highest doped *p*-type wafers where the c-Si surface is no longer in inversion for low injection levels and thus the double-diode model is no longer valid.

3.4. Determination of interface recombination parameters: interface recombination center density, field-effect passivation

Dauwe [Dau04] successfully proposes to model SiN_x/c-Si interface recombination by the extended SRH formalism including (the same as Kerr) recombination within a surface space-charge region. Dauwe models this recombination in the space-charge region with a SRH defect level having a lifetime of typically 1 µs, contrariwise to Kerr that supposes recombination in the space-charge region to be proportional to $e^{qV/(n_{02}kT)}$ (Sec. 3.3.3).

The similar $\tau_{eff,m}(\Delta n)$-dependance at SiO_2/c-Si and SiN_x/c-Si interfaces (compare Fig. 3.11 and Fig. 3.13) is thus due to different passivation properties: the injection level-dependance of SiO_2/c-Si $\tau_{eff}(\Delta n)$-curves results mostly from asymmetrical capture cross-sections while the injection level-dependance of SiN_x/c-Si $\tau_{eff}(\Delta n)$-curves is due a high fixed positive charge in combination with non-negligible recombination within the space-charge-region.

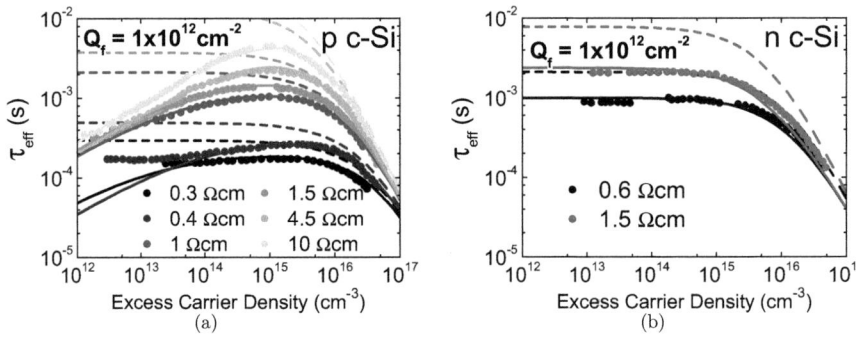

Figure 3.13: *Dots show measured effective lifetimes of Kerr's optimized stoichiometric SiN films passivating a) p- and b) n-type wafers of different doping levels [KC02a]. The upper lifetime limit as predicted by Kerr's parametrization is indicated by the dashed lines [KC]. Full lines show fits to the experimental data with the double-diode recombination model (Sec. 3.3.3) assuming that a high fixed positive charge density results in a) inversion of the p-type surface (accompanied by equal electron and hole density and therefore maximal SRH recombination in the space-charge region) and b) accumulation of the n-type surface even under illumination (much higher electron than hole density everywhere in the space-charge region and thus minimized recombination).*

3.5 Novel model for a-Si:H/c-Si interface recombination based on the amphoteric nature of silicon dangling bonds

The standard SRH recombination model is based on interface defects having two possible charge states. But c-Si surface defects were identified as dangling bonds and therefore we propose in this section to model heterostructure interface recombination by amphoteric defects, i.e. dangling bonds possessing three possible charge states. For this, we first introduce an existing recombination model for bulk a-Si:H based on the amphoteric nature of Si dangling bonds (3.5.2). This amphoteric recombination model is then applied for the first time to heterostructure interface recombination, more precisely to a-Si:H/c-Si interface recombination (3.5.3). Differences and similarities in comparison to the standard SRH recombination model are pointed out.

3.5.1 Introduction

The first Si solar cell passivation using a-Si:H was made in 1979 by Pankove et al. [PT79]. His reverse-biased *pn* junctions yielded two orders of magnitude lower leakage currents when passivated with a-Si:H instead of SiO_2.

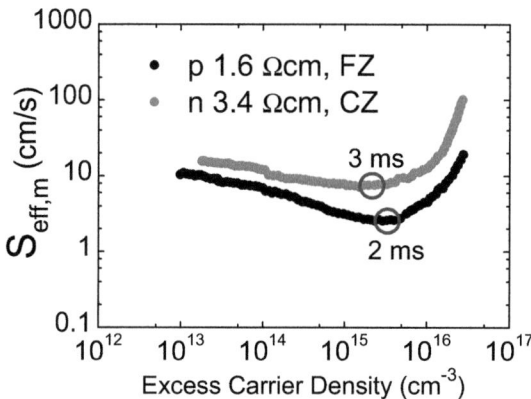

Figure 3.14: *Injection level dependent $S_{eff,m}$ of a 1.6 Ωcm FZ p-Si and a 3.4 Ωcm CZ n-Si wafer measured by Dauwe when passivated by $\sim$ 60 nm thick a-Si:H layers on both surfaces [Dau04].*

3.5. Novel model for a-Si:H/c-Si interface recombination based on the amphoteric nature of silicon dangling bonds

Dauwe et al. [DSH02] first achieved outstanding low surface recombination when passivating p- and n-type Si wafers with a-Si:H as shown in the $S_{eff,m}(\Delta n)$-plot in Fig. 3.14, where corresponding maximal lifetimes are indicated. The achieved passivation performance is comparable with the record SiO_2 values of Kerr (Fig. 3.11).

Subsequently, the a-Si:H/c-Si interface recombination's injection level-dependance has been modeled by Garin et al. [GRB+05] for the case of n a-Si:H / i a-Si:H / c-Si and p a-Si:H / i a-Si:H / c-Si heterostructures, by using the extended SRH formalism assuming a single level defect state in the middle of the c-Si's bandgap with equal electron and hole capture cross-sections. Vetter et al. [VMF+07] use the same model as Garin et al. for their amorphous carbide Si (a-SiC$_x$:H) passivation layers yielding measured effective lifetimes higher than 1 ms when deposited on 3.3 Ωcm p-type and 1.5 Ωcm n-type c-Si. They identify a field-effect passivation mechanism with a fixed positive charge for p-type and a negative one for n-type c-Si. In fact, when passivating c-Si by amorphous Si, the microscopic interface is different from the one at the SiO_2/c-Si and the SiN_x/c-Si interface. Because in the case of a-Si:H there are no other atoms than Si involved, all interface defects are of intrinsic nature, i.e. $Si \equiv Si\cdot$ dangling bonds forming states around midgap (named P_{b0} states for SiO_2 (Sec. 3.4.2.1)) and stretched Si−Si bonds leading to bandtail states. In addition, a-Si:H is, with its bandgap of ~ 1.75 eV, not a comparable insulator to SiO_2 and SiN_x, and therefore band offsets between a-Si:H and c-Si have to be considered.

Already in 1985 Biegelsen et al. [BJS+85] pointed out the similarity between the native defects at the SiO_2/c-Si interface and bulk a-Si:H, that is, these defect's amphoteric nature. In contrast to bulk c-Si defects, which possess 2 different charge conditions, Si dangling bond states have 3 different charge conditions, i.e. they are amphoteric. Hence, we extend a model previously established for amphoteric bulk a-Si:H recombination [VJ86, HSS92] to the description of the surface recombination through dangling bonds.

3.5.2 a-Si:H bulk recombination

To analytically resolve the injection level-dependance of the recombination rate R [cm^{-3}s^{-1}], one needs a closed-form expression. Shockley, Read and Hall (SRH) have basically considered one discrete recombination center in the gap, having two charge conditions. Taylor and Simmons extended this SRH formalism to the case of a continuous distribution of states within the gap [ST71], as present in disordered semiconductors. They have con-

3.5. Novel model for a-Si:H/c-Si interface recombination based on the amphoteric nature of silicon dangling bonds

cluded that, as long as the capture cross-sections of the electronic states do not vary as a function of their energy level, and as long as one can neglect thermal re-emission from them, the general form for R is similar to the one of a discrete recombination center level. However, in order to discriminate between the two roles that electronic states within the gap can play, i.e. they can act as trapping or as recombination centers, Taylor and Simmons introduced the notion of demarcation levels. The positions of these demarcation levels are defined by the equal probabilities of thermal emission and free carrier capture of an electronic state and depend on the generation rate G [cm^{-3}] (visualization in Fig. 3.15). Electronic states lying in-between the two demarcation levels act as recombination centers with negligible re-emission probability compared to the capture probabilities. Electronic states lying in-between demarcation level and band edge act as traps with non-negligible thermal re-emission probabilities. These demarcation levels, denoted E_{tn} for electrons and E_{tp} [eV] for holes depend on the generation rate through the free carrier densities $n_f = n_0 + \Delta n$ and $p_f = p_0 + \Delta n$ [cm^{-3}]:

$$E_{tn} = E_{Fn} + kT \ln(\frac{\sigma_p p_f + \sigma_n n_f}{\sigma_n n_f}),$$

$$E_{tp} = E_{Fp} - kT \ln(\frac{\sigma_n n_f + \sigma_p p_f}{\sigma_p p_f}), \qquad (3.29)$$

where E_{Fn} and E_{Fp} [eV] denote the electron and hole quasi-Fermi levels. The quasi-Fermi level concept is used in the description of non-equilibrium free excess carrier density distributions, when the Fermi-Dirac occupation function can no longer be used to describe the probability of occupation of electronic states:

$$E_{Fn} = -q\phi_n = kT \ln \frac{n_f}{n_i},$$

$$E_{Fp} = -q\phi_p = -kT \ln \frac{p_f}{n_i}. \qquad (3.30)$$

Note that besides n_f, also n_0, Δn and n_i are again (as defined in Sec. 3.2 and 3.3.1) the thermal equilibrium free carrier density, respectively the photogenerated excess carrier density and the intrinsic carrier density, but their values and the approximations made to find them are different for an amorphous than for a crystalline semiconductor (Fig. 3.22 and related discussions).

In bulk a-Si:H, monomolecular recombination via dangling bond (DB) states is the dominant mechanism for recombination at room temperature

3.5. Novel model for a-Si:H/c-Si interface recombination based on the amphoteric nature of silicon dangling bonds

and under low to medium illumination levels [Cra84]. The amphoteric DB states have three different charge conditions, i.e. their occupation by zero, one or two electrons leads to three different charge states of the Si_3 sites. They can be:

- either positively charged when not occupied by electrons (D^+) (trivalent bonded Si atom, Si_3^+),

- or neutral when occupied by one electron (D^0) (trivalent bonded Si atom with a singly occupied bond, Si_3^0),

- or negatively charged when occupied by two electrons (D^-) (trivalent bonded Si atom with two electrons on the dangling bond, Si_3^-). In this case, the energy level is shifted by an amount E_U (Fig. 3.15).

The correlation energy E_U [eV] is the energy required to place a second electron in the same Si_3 orbital, and thus the energy difference between the transition levels D^+/D^0 and D^0/D^-. In a-Si:H, E_U is positive because of the repulsive energy of having two electrons localized at the same dangling bond site, with a value of 0.3 ± 0.1 eV [PGR+84]. Assuming medium illumination level conditions, Hubin et al. [HSS92] found a closed-form expression for R where only the total density of dangling bonds (N_{DB} [cm^{-3}]) appears, that is, we do not need to know the shape of the continuous distribution of DB states.

Fig. 3.15 shows the typical single-electron representation of the distribution of recombination centers in a-Si:H. In the presented recombination model, only transitions of free carriers to localized dangling bond states are considered. Direct recombination between free carriers (band to band) and transitions between localized states are neglected. The demarcation levels E_{tn} for electrons and E_{tp} for holes are the quasi-Fermi levels for traps, and states in the energy interval $[E_{tn}, E_{tp}]$ act therefore as recombination centers and not as traps. By definition, thermal emission processes from such recombination centers can be neglected if the generation rate G is high enough for the demarcation levels to lie outside the midgap DB distribution (Fig. 3.15) [ST71]. Contrariwise, recombination in band tails can be neglected until G becomes high enough to push the demarcation levels very near the respective bands and thus into the bandtail states letting them act as recombination centers instead of traps (again Fig. 3.15).

3.5. Novel model for a-Si:H/c-Si interface recombination based on the amphoteric nature of silicon dangling bonds

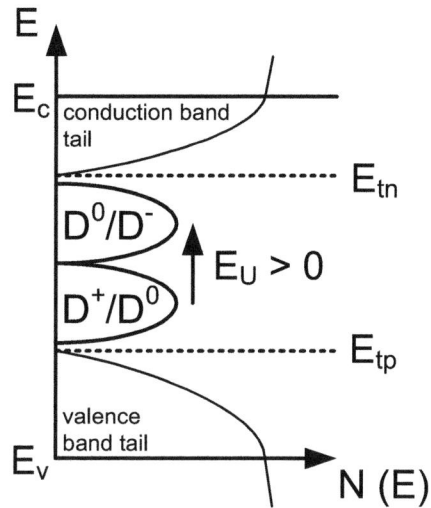

Figure 3.15: *Single-electron representation of a continuous distribution of amphoteric recombination centers (density of states $N(E)$). When unoccupied, the Si dangling bond (i.e. recombination center) is positively charged (D^+), when occupied by one electron, the recombination center is neutral (D^0). These two charge conditions are represented here at the same energy level. When occupied by two electrons, the recombination center is negatively charged (D^-), and if the correlation energy E_U is positive, it is represented as shifted upwards by E_U (as sketched here). E_{tn} and E_{tp} are the demarcation levels, i.e. the quasi-Fermi level for traps (see text) whose position depends on the generation rate G.*

For medium illumination levels this leads thus to the microscopic picture of DB recombination steps shown in Fig. 3.16 where two parallel recombination paths exist. Both consist of two successive capture events:

1. hole capture on a D^0 which changes the D^0 into a D^+ ($D^0 + h \rightarrow D^+$, capture rate r_p^0) followed by electron capture on this D^+ turning it again into a D^0 ($D^+ + e \rightarrow D^0$, capture rate r_n^+),

2. electron capture on a D^0 which changes the D^0 into a D^- ($D^0 + e \rightarrow D^-$, capture rate r_n^0) followed by hole capture on this D^- turning it again into a D^0 ($D^- + h \rightarrow D^0$, capture rate r_p^-).

3.5. Novel model for a-Si:H/c-Si interface recombination based on the amphoteric nature of silicon dangling bonds

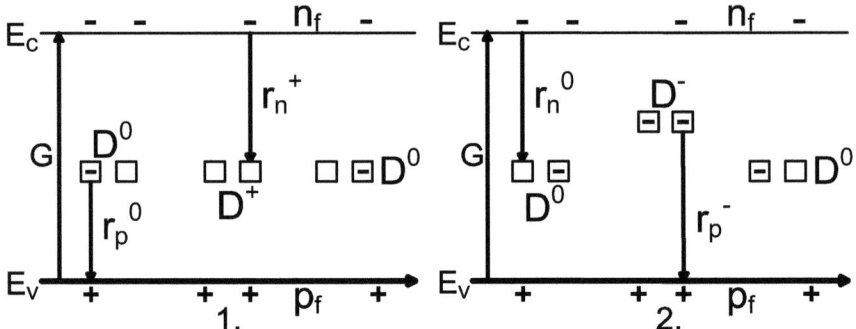

Figure 3.16: *Successive capture events leading to electron-hole recombination through D^0, i.e. when most recombination centers are in the neutral state (positive correlation energy E_U). Two paths co-exist (denoted by 1. and 2.). Thermal emission processes are neglected. n_f and p_f are the free carrier densities, D^0, D^+ and D^- are the neutral, positively, and negatively charged conditions of the dangling bonds and r_p^0, r_n^+, r_n^0 and r_p^- are the corresponding capture rates.*

The capture rates [cm^{-3}s^{-1}] are given by (see for example [Str84]):

$$\begin{aligned} r_p^0 &= v_{th}\sigma_p^0 p_f f_{DB}^0 N_{DB}, \\ r_n^+ &= v_{th}\sigma_n^+ n_f f_{DB}^+ N_{DB}, \\ r_n^0 &= v_{th}\sigma_n^0 n_f f_{DB}^0 N_{DB}, \\ r_p^- &= v_{th}\sigma_p^- p_f f_{DB}^- N_{DB}, \end{aligned} \quad (3.31)$$

where v_{th} [cm/s] is the thermal velocity, N_{DB} [cm^{-3}] is the total density of dangling bonds, σ_n^0 and σ_p^0 [cm^2] are the capture cross-sections of the neutral states and σ_n^+ and σ_p^- [cm^2] are the capture cross-sections of the charged states. f_{DB}^0 [ ] is the probability that a DB is neutral, f_{DB}^+ [ ] is the probability that a DB is charged positively and f_{DB}^- [ ] is the probability that a DB is charged negatively, where $f_{DB}^0 + f_{DB}^+ + f_{DB}^- = 1$. When neglecting thermal emission processes, steady-state is attained for:

$$\begin{aligned} r_p^0 &= r_n^+ \quad \text{and} \\ r_n^0 &= r_p^-. \end{aligned} \quad (3.32)$$

3.5. Novel model for a-Si:H/c-Si interface recombination based on the amphoteric nature of silicon dangling bonds

From Eqs. 3.31 and 3.32 the DB occupation functions can be found:

$$f_{DB}^+ = \frac{\sigma_p^- \sigma_p^0 p_f^2}{\sigma_p^- \sigma_p^0 p_f^2 + \sigma_n^+ \sigma_p^- n_f p_f + \sigma_n^0 \sigma_n^+ n_f^2},$$

$$f_{DB}^- = \frac{\sigma_n^+ \sigma_n^0 n_f^2}{\sigma_p^- \sigma_p^0 p_f^2 + \sigma_n^+ \sigma_p^- n_f p_f + \sigma_n^0 \sigma_n^+ n_f^2},$$

$$f_{DB}^0 = 1 - f_{DB}^+ - f_{DB}^-. \quad (3.33)$$

Fig. 3.17 shows their plot as a function of the free carrier density ratio n_f/p_f for different capture cross-section ratios. The maximum of f_{DB}^0 is reached at

$$f_{DB,max}^0 = \sqrt{\frac{\sigma_p^0 \sigma_p^-}{\sigma_n^0 \sigma_n^+}} = (\frac{\sigma_n^0}{\sigma_p^0})^{-1} \times \sqrt{\frac{\sigma_p^-/\sigma_p^0}{\sigma_n^+/\sigma_n^0}}, \quad (3.34)$$

and 3 zones can be distinguished:

1. $n_f/p_f \approx f_{DB,max}^0$: $f_{DB}^0 \approx 1 \gg f_{DB}^+ \approx f_{DB}^-$,

2. $n_f/p_f \gg f_{DB,max}^0$: $f_{DB}^0 + f_{DB}^- \approx 1$ and $f_{DB}^+ \approx 0$,

3. $n_f/p_f \ll f_{DB,max}^0$: $f_{DB}^0 + f_{DB}^+ \approx 1$ and $f_{DB}^- \approx 0$.

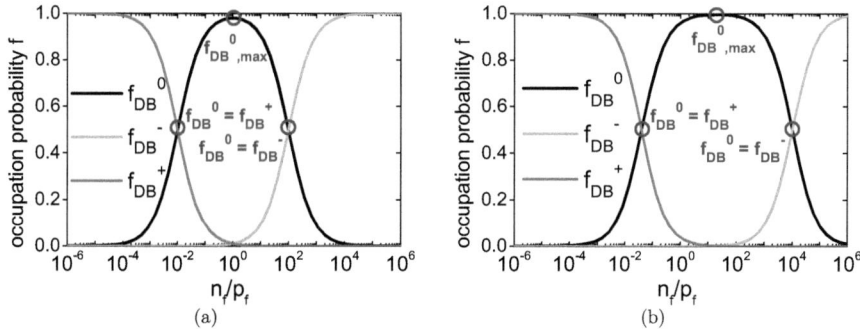

Figure 3.17: *DB occupation probabilities f_{DB}^0, f_{DB}^+ and f_{DB}^- as a function of the free carrier density ratio n_f/p_f for a)* $\frac{\sigma_n^0}{\sigma_p^0} = 1 / \frac{\sigma_n^+}{\sigma_p^0} = \frac{\sigma_p^-}{\sigma_n^0} = 100$ *and b)* $\frac{\sigma_n^0}{\sigma_p^0} = \frac{1}{20} / \frac{\sigma_n^+}{\sigma_p^0} = \frac{\sigma_p^-}{\sigma_n^0} = 500$. f_{DB}^0, f_{DB}^+ *and* f_{DB}^- *are independent of E_U if the capture cross-sections are energy independent.*

3.5. Novel model for a-Si:H/c-Si interface recombination based on the amphoteric nature of silicon dangling bonds

The total dangling bond density is given by

$$\rho_{DB}^0 + \rho_{DB}^+ + \rho_{DB}^- = f_{DB}^0 N_{DB} + f_{DB}^+ N_{DB} + f_{DB}^- N_{DB} = N_{DB}, \quad (3.35)$$

where ρ_{DB}^0, ρ_{DB}^+ and ρ_{DB}^- [cm^{-3}] denote the densities of the differently charged DBs. Among both recombination paths in Fig. 3.16, recombination is limited by the less probable capture event (the smaller capture rate). Under steady-state conditions, if $r_p^0 < r_n^+$, most DBs are in the neutral state, and if additionally $r_n^0 < r_p^-$ DBs in the negatively charged state are rare as well. If, in contrast, $r_p^0 \gg r_n^+$ and $r_n^0 \gg r_p^-$, most DBs are either in one or the other charge state and the resulting microscopic picture is more SRH-like. However, in the microscopic picture of SRH there is only one recombination path through a recombination level having two charge conditions.

Three hypotheses have to be made to find the closed-form expression for R:

1. The steady-state condition is fulfilled independently at each energy level, i.e. the total electron capture rate equals the total hole capture rate, where this rate is the DB recombination rate R_{DB} [cm^{-3}s^{-1}] given by

$$R_{DB} = r_n^0 + r_n^+ = r_p^- + r_p^0. \quad (3.36)$$

2. The illumination level is high enough, so that the demarcation levels for electrons (E_{tn}) and holes (E_{tp}) lie outside the distribution of DB states (Fig. 3.15).

3. The capture cross-sections of the DB states are independent of the energy level.

Under these three hypotheses, the calculation of the recombination rate can be reduced to the case of a discrete recombination level with three charge conditions [HSS92]. From Eqs. 3.31, 3.33 and 3.36, the resulting recombination rate can be written in terms of the capture cross-sections, the free carrier densities and the total DB density as

$$R_{DB}(n_f, p_f) = \frac{n_f \sigma_n^0 + p_f \sigma_p^0}{\frac{p_f}{n_f}\frac{\sigma_p^0}{\sigma_n^+} + 1 + \frac{n_f}{p_f}\frac{\sigma_n^0}{\sigma_p^-}} v_{th} N_{DB}. \quad (3.37)$$

Note that this recombination rate does not depend on the correlation energy E_U and the particular shape of the density of states of the dangling bonds $N(E)$ [cm^{-3}eV^{-1}] (Fig. 3.15), due to the three hypotheses made for

3.5. Novel model for a-Si:H/c-Si interface recombination based on the amphoteric nature of silicon dangling bonds

the calculation. The correlation energy E_U is implicitly contained in the charged to neutral capture cross-section ratios $\frac{\sigma_p^-}{\sigma_n^0}$ and $\frac{\sigma_n^+}{\sigma_p^0}$.

For $n_f \approx p_f$, the larger the ratio $\frac{\sigma^\pm}{\sigma^0}$, the greater the number of DBs in the neutral state, and the more pronounced the difference between R_{DB} and R_{SRH}: when the two conditions $\frac{\sigma_n^+}{\sigma_p^0} \gg p_f/n_f$ and $\frac{\sigma_p^-}{\sigma_n^0} \gg n_f/p_f$ are fulfilled simultaneously, the expression for R_{DB} reduces to:

$$R_{DB}(n_f, p_f) = (n_f \sigma_n^0 + p_f \sigma_p^0) v_{th} N_{DB}, \qquad (3.38)$$

which means that the recombination rate is limited by the larger free carrier density which is opposite to the SRH case. This range will be called the "DB_i-domain". Conversely, for the SRH expression the recombination is limited by the lowest free carrier density [SR52, Hal52]:

$$R_{SRH}(n_f, p_f) = \frac{n_f p_f}{\frac{n_f}{\sigma_p} + \frac{p_f}{\sigma_n}} v_{th} N_t = \left(\frac{1}{\sigma_p p_f} + \frac{1}{\sigma_n n_f} \right)^{-1} v_{th} N_t. \qquad (3.39)$$

For extrinsic, sufficiently doped a-Si:H, one type of charged dangling bonds is dominant (e.g. in n-type doped a-Si:H, most DBs are negatively charged), and finally Eq. 3.37 reduces to a SRH-like rate, and consequently a microscopic picture similar to SRH recombination is reached. For example, taking n-type material, where $n_f \gg p_f$, the expressions for R are similar:

- By DB recombination, Eq. 3.37 with $n_f/p_f \gg \frac{\sigma_p^-}{\sigma_n^0}$ reduces to $R_{DB} = p_f \sigma_p^- v_{th} N_{DB}$, i.e. recombination occurs through path 2 in Fig. 3.16 (as $r_p^0 < r_n^+$ and $r_n^0 < r_p^-$, if $n_f \gg p_f$ then $r_n^0 \gg r_p^0$ and recombination path 1 disappears),

- whereas by SRH recombination, Eq. 3.39 with $n_f/p_f \gg \frac{\sigma_p}{\sigma_n}$ reduces to $R_{SRH} = p_f \sigma_p v_{th} N_t$.

Therefore, this range of recombination will be called the "SRH-domain".

The surface plot of R_{DB} as a function of n_f and p_f (Eq. 3.37) is shown in Fig. 3.18. The ratios of the capture cross-sections are set to the values given in Chap. 4 which allow the best fit of the experimental data with our model: the ratio of neutral electron to hole capture cross-section $\frac{\sigma_n^0}{\sigma_p^0}$ is set to $\frac{1}{20}$ and the ratio of charged to neutral capture cross-section is set to $\frac{\sigma_n^+}{\sigma_n^0} = \frac{\sigma_p^-}{\sigma_p^0} = 500$.

3.5. Novel model for a-Si:H/c-Si interface recombination based on the amphoteric nature of silicon dangling bonds

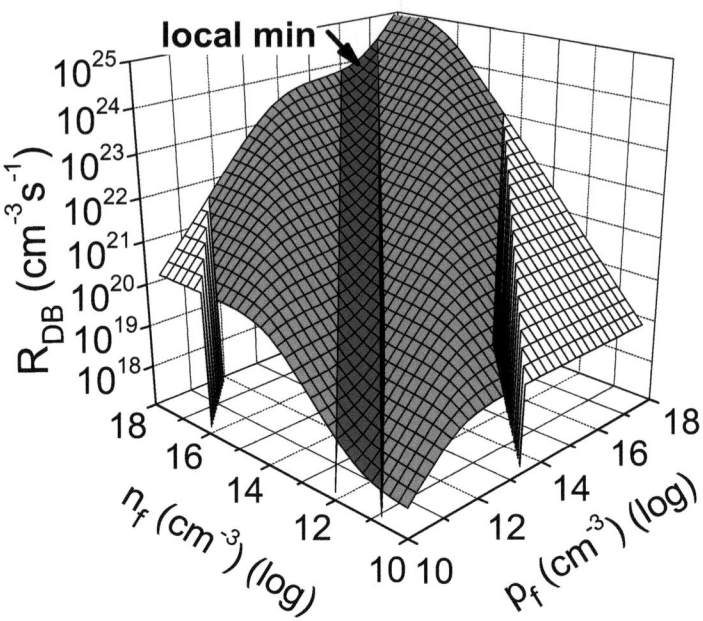

Figure 3.18: *Recombination rate R_{DB} in a-Si:H as a function of the free carrier densities n_f and p_f. In this plot, capture cross-section ratios are set to $\frac{\sigma_n^0}{\sigma_p^0} = \frac{1}{20}$ and $\frac{\sigma_n^+}{\sigma_n^0} = \frac{\sigma_p^-}{\sigma_p^0} = 500$. Over the two (light and dark) grey surfaces, the recombination cannot be described by a SRH-like recombination rate, it has to be described by Eq. 3.37 and is therefore called the "DB-domain". In its center, the dark grey surface denotes the "DB_i-domain", where Eq. 3.37 can be simplified to Eq. 3.38. Its position is mainly given by the neutral capture cross-section ratio $\frac{\sigma_n^0}{\sigma_p^0}$ for similarily charged to neutral capture cross-section ratios (Eq. 3.40). Within the width of the total grey surface determined by the charged to neutral capture cross-section ratios, recombination is dominated by majority carriers and thus opposite to the common SRH recombination. The white surface is the "SRH-domain", where the recombination rate R_{DB} reduces to a R_{SRH}-like rate, i.e. the recombination process is limited by minority carriers.*

The chosen value for the dangling bond density $N_{DB} = 5 \times 10^{15}$ cm^{-3} is within the range of the commonly accepted values for device grade a-Si:H ($N_{DB} = 1 \times 10^{15} - 1 \times 10^{16}$ cm^{-3}) [LSK06, ENP03]. The additionally used values are $\sigma_p^0 = 1 \times 10^{-16}$ cm^2 and $v_{th} = 2 \times 10^7$ cm/s. They are further discussed in Sec. 3.5.3.

3.5. Novel model for a-Si:H/c-Si interface recombination based on the amphoteric nature of silicon dangling bonds

The following general trends can be noted in Fig. 3.18: if one aims at minimizing recombination in undoped material (i.e. material in which $n_f \approx p_f$), one recognizes in Fig. 3.18, that there is a local minimum at high R_{DB} (dark grey surface), occurring when the denominator in Eq. 3.37 is maximal, i.e.

$$n_f/p_f = (\frac{\sigma_n^0}{\sigma_p^0})^{-1} \times \sqrt{\frac{\sigma_p^-/\sigma_p^0}{\sigma_n^+/\sigma_n^0}}, \qquad (3.40)$$

that equals $f^0_{DB,max}$ from Eq. 3.34 (Fig. 3.17). This local minimum of R_{DB}, mainly given by the ratio of the neutral capture cross-sections is surrounded by two local maxima found by setting the first or the last term in the denominator of Eq. 3.37 equal to 1 and is given by $n_f/p_f = (\frac{\sigma_n^0}{\sigma_p^0})^{-1} \times (\frac{\sigma_n^+}{\sigma_n^0})^{-1}$ and $n_f/p_f = (\frac{\sigma_n^0}{\sigma_p^0})^{-1} \times \frac{\sigma_p^-}{\sigma_p^0}$ (c.f. Fig. 3.17, where $f^0_{DB} = f^+_{DB}$ and $f^0_{DB} = f^-_{DB}$). The width of the range between these two local maxima is thus determined by the ratio of charged to neutral capture cross-sections. Within this whole range, recombination cannot be described by a SRH-like recombination rate, it has to be described by Eq. 3.37 (grey surface in Fig. 3.18) and is thus fundamentally different from recombination through a recombination center having only two charge states. The absolute minimum value of R_{DB} can be reached when $n_f \neq p_f \gg 1$ or $\ll 1$, i.e. by field-effect, where by strongly reducing the density of one carrier type we reach the same microscopic picture as for extrinsic, doped a-Si:H, and thus as SRH recombination (white surface in Fig. 3.18). The positions of the local minimum, the local maxima and the onset of SRH-like recombination is thus solely given by the capture cross-section ratios $\frac{\sigma_n^0}{\sigma_p^0}$, $\frac{\sigma_n^+}{\sigma_n^0}$ and $\frac{\sigma_p^-}{\sigma_p^0}$.

In contrast, for a recombination center having two possible charge states, the occupation probabilities f^0_{SRH} [ ] for the recombination center to be neutral and f^-_{SRH} for the recombination center to be negatively charged are given by [SR52]:

$$f^0_{SRH} = \frac{\sigma_p p_f + \sigma_n n_1}{\sigma_n(n_f + n_1) + \sigma_p(p_f + p_1)},$$

$$f^-_{SRH} = \frac{\sigma_n n_f + \sigma_p p_1}{\sigma_n(n_f + n_1) + \sigma_p(p_f + p_1)},$$

where $n_1 \equiv n_i e^{(E_t - E_i)/kT}$ and $p_1 \equiv n_i e^{-(E_t - E_i)/kT}$ \qquad (3.41)

3.5. Novel model for a-Si:H/c-Si interface recombination based on the amphoteric nature of silicon dangling bonds

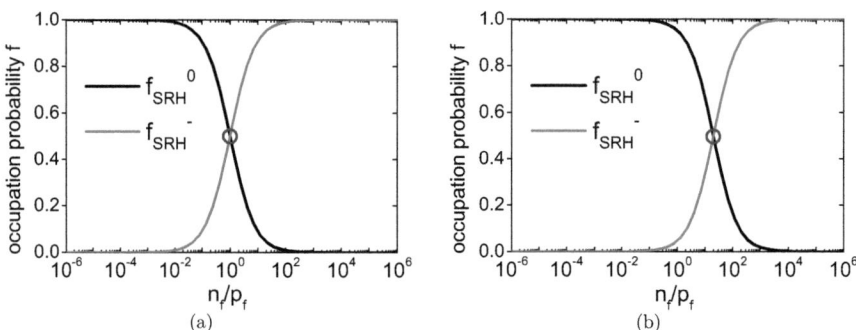

Figure 3.19: *SRH occupation probabilities f^0_{SRH} and f^-_{SRH} as a function of the free carrier density ratio n_f/p_f for a) $\frac{\sigma_n}{\sigma_p} = 1$ and b) $\frac{\sigma_n}{\sigma_p} = \frac{1}{20}$.*

Fig. 3.19 shows their plot as a function of the free carrier density ratio n_f/p_f. Recombination through such a recombination center is given by Eq. 3.39 and plotted as a function of n_f and p_f in Fig. 3.20 for two different capture cross-section ratios together with the surface plot of R_{DB} from Eq. 3.37. Maximal recombination is given by maximizing Eq. 3.39's denominator, i.e. $n_f/p_f = \left(\frac{\sigma_n}{\sigma_p}\right)^{-1}$.

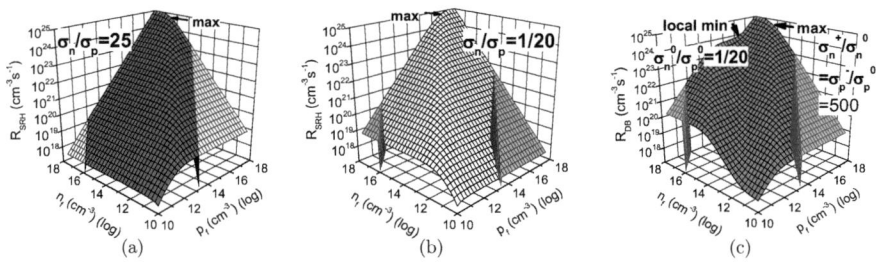

Figure 3.20: *Comparison of the standard SRH and the amphoteric bulk recombination rates R_{SRH} a),b) and R_{DB} c). The capture cross-section ratios are set to a) $\frac{\sigma_n}{\sigma_p} = \frac{\sigma_n^0}{\sigma_p^0} \times \frac{\sigma_n^+}{\sigma_n^0} = 25$, b) $\frac{\sigma_n}{\sigma_p} = \frac{\sigma_n^0}{\sigma_p^0} = \frac{1}{20}$ and c) $\frac{\sigma_n^0}{\sigma_p^0} = \frac{1}{20} / \frac{\sigma_n^+}{\sigma_n^0} = \frac{\sigma_p^-}{\sigma_p^0} = 500$. With the given capture cross-section ratios, R_{DB} is over the dark gray range more similar to R_{SRH} in a), while only over the light gray range it is more similar to R_{SRH} in b) having an equal neutral capture cross-section ratio.*

3.5. Novel model for a-Si:H/c-Si interface recombination based on the amphoteric nature of silicon dangling bonds

Only around a sharp maximum recombination is majority carrier density limited. Minority carrier density limited recombination is predominant for all other free carrier densities.

The fillings in the surface plots of Fig. 3.20 illustrate the different similar regions for the selected capture cross-section ratios. On one hand, for equal neutral capture cross-section ratios (compare Figs. 3.20(b) and 3.20(c)), i.e. $\frac{\sigma_n}{\sigma_p} = \frac{\sigma_n^0}{\sigma_p^0}$, R_{SRH} has its maximum while R_{DB} has a local minimum. Only for $n_f \neq p_f \gg 1$ or $\ll 1$, i.e. the light gray surface, R_{SRH} equals R_{DB}. On the other hand, for a neutral capture cross-section ratio given by $\frac{\sigma_n}{\sigma_p} = \frac{\sigma_n^0}{\sigma_p^0} \times \frac{\sigma_n^+}{\sigma_n^-}$ (compare Figs. 3.20(a) and 3.20(c)), the absolute maximum of R_{SRH} coincides with the one of R_{DB}. Over the dark gray range, R_{SRH} is similar to R_{DB}.

3.5.3 Extension to a-Si:H/c-Si interface recombination

Using the same description of recombination via a-Si:H/c-Si interface dangling bond states as for bulk a-Si:H, the DB interface recombination rate U_{DB} [cm^{-2}s^{-1}] is given in analogy to Eq. 3.37 and Eq. 3.12 by:

$$U_{DB}(n_s, p_s) = \frac{n_s \sigma_n^0 + p_s \sigma_p^0}{\frac{p_s}{n_s} \frac{\sigma_p^0}{\sigma_n^+} + 1 + \frac{n_s}{p_s} \frac{\sigma_n^0}{\sigma_p^-}} v_{th} N_s, \qquad (3.42)$$

where the three-dimensional bulk DB density N_{DB} [cm^{-3}] reduces to a two-dimensional interface state density N_s [cm^{-2}]. For flatband conditions, recombination at the c-Si surface is described by the surface recombination velocity S_{DB} [cm/s] (Sec. 3.3.1):

$$S_{DB}(\Delta n) = \frac{1}{\Delta n} \frac{(n_0 + \Delta n)\sigma_n^0 + (p_0 + \Delta n)\sigma_p^0}{\frac{(p_0+\Delta n)}{(n_0+\Delta n)} \frac{\sigma_p^0}{\sigma_n^+} + 1 + \frac{(n_0+\Delta n)}{(p_0+\Delta n)} \frac{\sigma_n^0}{\sigma_p^-}} v_{th} N_s. \qquad (3.43)$$

Fig. 3.21(c) visualizes $U_{DB} = f(n_s, p_s)$ from Eq. 3.42 with the same capture cross-sections as used for Fig. 3.18. When varying experimentally, as in our Sinton lifetime measurement, Δn from 1×10^{17} cm^{-3} to 1×10^{12} cm^{-3}, the variations of n_s and p_s are given by the wafer doping level solely, as $n_s = n_0 + \Delta n$ and $p_s = p_0 + \Delta n$. The n_s, p_s-projections for the three different bulk doping levels shown are thus identical in the surface

85

3.5. Novel model for a-Si:H/c-Si interface recombination based on the amphoteric nature of silicon dangling bonds

plots of Fig. 3.21, where 3.21(a) and 3.21(b) additionally show U_{SRH} with the same two capture cross-section ratios as in Fig. 3.20. But depending on the assumed surface recombination rate U_s, the resulting $S(\Delta n)$-curves shown in Fig. 3.21(d) differ significantly, as becomes obvious from the corresponding trajectories on the surface plots in Fig. 3.21.

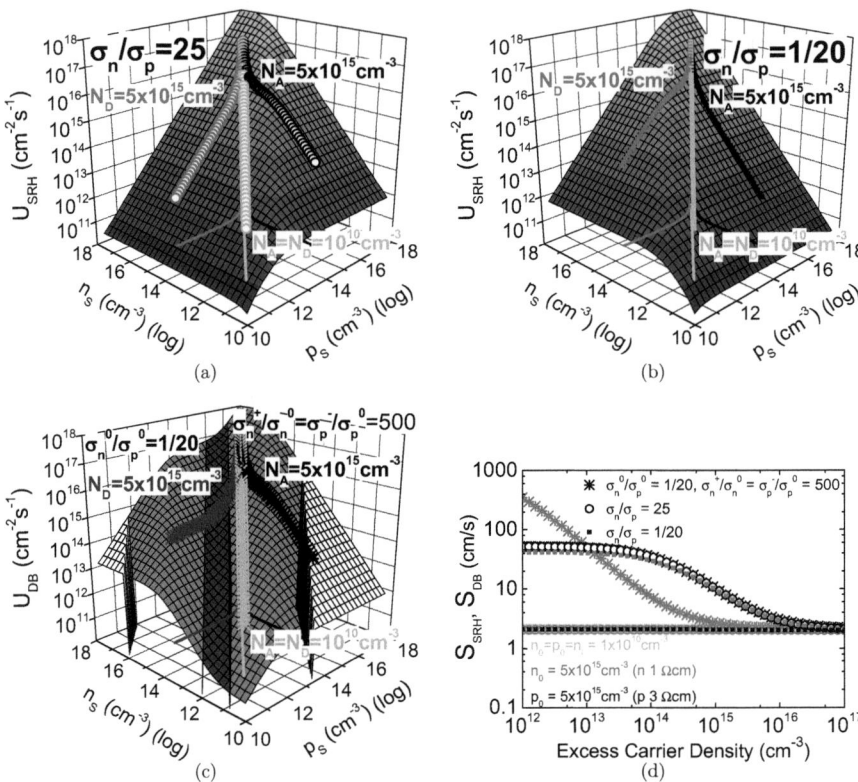

Figure 3.21: *Comparison of the standard SRH and the amphoteric surface recombination rates U_{SRH} a),b) and U_{DB} c). The capture cross-section ratios are set according to Fig. 3.20. While the n_s, p_s-projections are identical (they depend only on the wafer doping and the injection level), the resulting trajectories on the surface plots and thus the $S(\Delta n)$-curves in Fig. d) depend significantly on the assumed surface recombination rate U_s.*

In analogy to the case of SRH recombination for which the model had to be extended for non-flatband conditions at the c-Si surface, the effec-

3.5. Novel model for a-Si:H/c-Si interface recombination based on the amphoteric nature of silicon dangling bonds

tive surface recombination velocity S_{eff} [cm/s] commonly describes two-dimensional recombination (Sec. 3.3.2):

$$S_{eff,DB}(\Delta n) = \frac{1}{\Delta n} \frac{n_s \sigma_n^0 + p_s \sigma_p^0}{\frac{p_s}{n_s}\frac{\sigma_p^0}{\sigma_n^+} + 1 + \frac{n_s}{p_s}\frac{\sigma_n^0}{\sigma_p^-}} v_{th} N_s. \qquad (3.44)$$

Again, the surface potential ψ_s determines the interface carrier densities n_s and p_s (Eq. 3.20). Assuming a unidirectional diffusion current flow from the c-Si into the a-Si:H used here (intrinsic or microdoped), the passivation layer is implicitly considered as a small bandgap insulator. The band offset between a-Si:H (bandgap of about 1.75 eV) and c-Si (bandgap of 1.12 eV) is not taken explicitly into account but it will intervene through means of an additional small surface band bending (see later on in this section). Neglecting additional recombination in the Si space-charge region, the concept used in the extended SRH formalism developed in Sec. 3.3.2 applies. In the following we will discuss (in analogy to Sec. 3.3.2) the influence of bulk c-Si doping and field-effect passivation on the injection level dependent surface recombination velocities and thus, finally on the experimentally accessible carrier lifetimes calculated with the amphoteric surface recombination rate.

In contrast to c-Si, in which the bulk charge neutrality condition is simply given by the free carrier and ionized acceptor and donor densities, the charge neutrality condition in bulk a-Si:H is given by [Sau92]

$$n_f + n_t + \rho_{DB}^- + \rho_A^- = p_f + p_t + \rho_{DB}^+ + \rho_D^+, \qquad (3.45)$$

that is the balance between:

- n_f [cm^{-3}]: free electrons in the conduction band,
- p_f [cm^{-3}]: free holes in the valence band,
- n_t [cm^{-3}]: electrons localized in the conduction bandtail,
- p_t [cm^{-3}]: holes localized in the valence bandtail,
- ρ_{DB}^- [cm^{-3}]: negatively charged DBs,
- ρ_{DB}^+ [cm^{-3}]: positively charged DBs,
- ρ_A^- [cm^{-3}]: ionized acceptors,
- and ρ_D^+ [cm^{-3}]: ionized donors.

3.5. Novel model for a-Si:H/c-Si interface recombination based on the amphoteric nature of silicon dangling bonds

The localized charge densities in the bandtails are higher than the density of photogenerated free carriers, i.e. $n_t \gg n_f$ and $p_t \gg p_f$ as illustrated in Fig. 3.22, and thus Eq. 3.45 reduces to

$$n_t + \rho_{DB}^- + \rho_A^- = p_t + \rho_{DB}^+ + \rho_D^+. \tag{3.46}$$

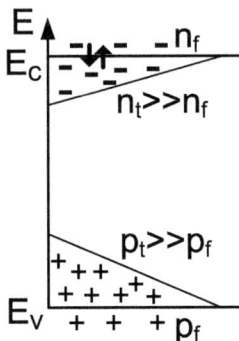

Figure 3.22: *Localized charge densities in bandtails and free carrier densities. The localized charge density in a bandtail depends only on the density of one free carrier type.*

Setting additionally the net bulk DB charge density $Q_{DB,bulk} = -\rho_{DB}^- + \rho_{DB}^+$ [cm^{-3}] and the density of localized bulk charge $Q_{t,bulk} = -n_t + p_t$ [cm^{-3}] and considering undoped material, we can rewrite the charge neutrality condition from Eq. 3.45 in the a-Si:H bulk as:

$$Q_{t,bulk} + Q_{DB,bulk} = 0. \tag{3.47}$$

Note that while recombination happens only through DBs, the bandtail states have to be considered for charge neutrality. At the interface to c-Si, Eq. 3.47 does not need to equal 0 as the net charge can be balanced within the c-Si space-charge region:

$$Q_t + Q_{DB} = Q_s = -Q_{Si}. \tag{3.48}$$

The density of localized interface charge Q_t [cm^{-2}] is given by the corresponding free carrier density (again Fig. 3.22) and the density of charged interface DBs Q_{DB} [cm^{-2}] depends additionally on the DB density N_s. If N_s is small, Eq. 3.48 reduces to

$$Q_s = Q_t. \tag{3.49}$$

3.5. Novel model for a-Si:H/c-Si interface recombination based on the amphoteric nature of silicon dangling bonds

Note that, independent of the sign of Q_t, the sign of Q_{DB} governing interface recombination depends solely on the occupation probabilities $f_{DB}^{+,-} = f(n_s, p_s)$ given in Eq. 3.33. Therefore the bandtails act as a charge reservoir while they do not participate in recombination. The charge stored in the a-Si:H near the interface in the dangling bonds $Q_{DB,bulk}$ and the bandtails $Q_{t,bulk}$ depends on the Si wafer's dark Fermi level, while the interface Q_{DB} varies depending on n_s, p_s such as given by ψ_s calculated from Eq. 3.24, where $Q_s = Q_{DB} + Q_t = -Q_{Si}$. Hence, in the following we set $Q_s = Q_{DB} + Q_t$ to an injection level independent value whose sign and magnitude is solely governed by the Si wafers doping density. This will be justified in the experimental part in Chap. 4.

The effects of the microscopic recombination model parameters (i.e. the capture cross-section ratios) were already discussed in Sec. 3.5.2. They will be set individually to an unique value for all a-Si:H/c-Si combinations later on. By setting $\frac{\sigma_n^+}{\sigma_n^0}$ equal to $\frac{\sigma_p^-}{\sigma_p^0}$ we can reduce the model's parameter number, in agreement with most data published in the literature [HSSP95]. $\sigma^{+,-}$ are generally assumed to be much larger than σ^0 [BWHS96]. The variation of the capture cross-section ratio $\frac{\sigma_n^0}{\sigma_p^0}$ modifies the symmetry of the surface plot in Fig. 3.18 and sets the local minimum of U_{DB} with respect to $n_s = p_s$. $\frac{\sigma_n^+}{\sigma_n^0} = \frac{\sigma_p^-}{\sigma_p^0}$ determines the width of the DB-domain and thus enlarges the range of the local minimum of U_{DB}. As can be seen readily from Eq. 3.44, $S_{eff,DB}$ is simply proportional to the interface recombination center density N_s representing the quality of the passivation. Equation 3.44 can be rewritten as

$$S_{eff,DB}(\Delta n) = \frac{1}{\Delta n} \frac{n_s \frac{\sigma_n^0}{\sigma_p^0} + p_s}{\frac{p_s}{n_s} \frac{\sigma_p^0}{\sigma_n^+} + 1 + \frac{n_s}{p_s} \frac{\sigma_n^0}{\sigma_p^-}} v_{th} N_s \sigma_p^0, \tag{3.50}$$

where σ_p^0 appears in the product $N_s \times \sigma_p^0$ and N_s is thus scaled with σ_p^0. A reasonable value of $\sigma_p^0 = 1 \times 10^{-16}$ cm^2 is chosen, as usual for neutral midgap states [MM04]. The thermal velocity is set to $v_{th} = 2 \times 10^7$ cm/s, i.e. to the one of c-Si, where $v_{th_n} \approx v_{th_p}$, as the effective masses of electrons and holes are only slightly different. The neutral electron and hole capture cross-sections σ_n^0 and σ_p^0 are generally assumed to be in the same range [HSSP95].

In order to illustrate the effect of the remaining recombination model parameter Q_s, we calculate $S_{eff,DB}$ as a function of the excess carrier density with the microscopic parameters set to the values that concur the most with our whole set of experimental data (Chap. 4), $\frac{\sigma_n^0}{\sigma_p^0} = \frac{1}{20}$ and $\frac{\sigma_n^+}{\sigma_n^0} = \frac{\sigma_p^-}{\sigma_p^0} = 500$.

3.5. Novel model for a-Si:H/c-Si interface recombination based on the amphoteric nature of silicon dangling bonds

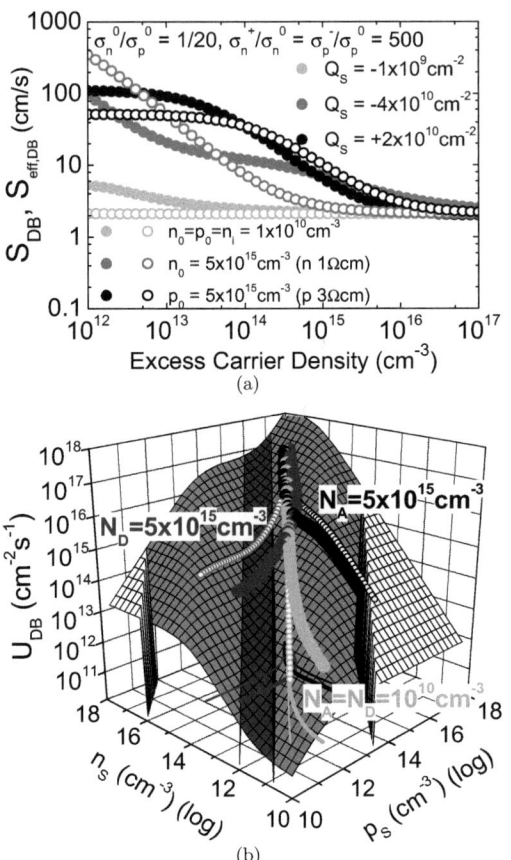

Figure 3.23: *Effect of surface band banding due to interface charge $Q_s = -Q_{Si}$ (full circles) compared to flatband conditions (open circles). a) Surface recombination velocities as a function of the bulk injection level calculated with the amphoteric recombination model for three differently doped wafers with three different surface charges Q_s. b) Visualization by means of surface charge based n_s, p_s-projection variation. The plotted trajectories on the surface recombination rate's surface plot are a representation of the $S(\Delta n)$-curves in Fig. a).*

Figure 3.23 shows again the case of three differently doped c-Si wafers passivated with intrinsic a-Si:H. Assuming a larger valence than conduction band offset (1.75 eV bandgap of a-Si:H and 1.12 eV bandgap of c-Si, usual

3.5. Novel model for a-Si:H/c-Si interface recombination based on the amphoteric nature of silicon dangling bonds

band alignment shown in Fig. 5.4), this would result in a diffusion of electrons from the intrinsic c-Si into the intrinsic a-Si:H and therefore a small negative surface charge of $Q_s = -1 \times 10^9$ cm^{-2} is assumed. For the same reasons, with the same p- and n-type c-Si doping level, a higher surface charge of $Q_s = -4 \times 10^{10}$ cm^{-2} is assumed for n-type than the one for p-type of $Q_s = +2 \times 10^{10}$ cm^{-2}.

The variations of U_{DB} as a function of n_s and p_s are visualized by means of the surface plot of U_{DB} in Fig. 3.23 for the three couples $n_0 = p_0 = n_i = 1 \times 10^{10}$ cm^{-3}/$Q_s = -1 \times 10^9$ cm^{-2}, $n_0 = 5 \times 10^{15}$ cm^{-3}/$Q_s = -4 \times 10^{10}$ cm^{-2} and $p_0 = 5 \times 10^{15}$ cm^{-3}/$Q_s = +2 \times 10^{10}$ cm^{-2}. From the replication of the flatband surface plot trajectories from Fig. 3.21(c) by means of the open circles and their white n_s, p_s-projections in Fig. 3.23(b), one sees that these specific induced charges only slightly modify the interface carrier densities and thus do not change the general trend of the trajectories on the surface plot and the $S(\Delta n)$-curves in Fig. 3.23.

In contrast to c-Si, the major effect of ionized doping impurities in a-Si:H is to modify the average state of charge of the DBs, which results in the simultaneous variation of the free carrier densities only for high a-Si:H doping impurity concentrations [Str91]. Low-level doping (called microdoping) varies the average state of charge of the DBs and, thus, Q_s without greatly influencing the total DB density N_{DB}. Heavier doping results in an increase in the dangling bond density N_{DB}, and no longer permits the variation of Q_s independently of N_{DB}. Thus, doped a-Si:H layers are only used in stack with i a-Si:H layers in this work. These configurations allow the modification of the average state of charge of the DBs by the electric field imposed in the i a-Si:H layers when fixing their outer surface potential. Q_s is, thus, representative of the magnitude of the field-effect on the passivation. Figure 3.24 shows the case of a lightly p-type doped c-Si wafer ($N_A = 1 \times 10^{14}$ cm^{-3}) passivated with a-Si:H of varying Q_s. For the purpose of comparison, the regions of the $S_{eff,DB}(\Delta n)$-curves in Fig. 3.24(a) highlighted with stars correspond to the SRH-domain in Fig. 3.24(b) where the amphoteric surface recombination rate in Eq. 3.42 reduces to a standard SRH-like rate (minority carrier density limited recombination). The rest of the curves belong to the DB-domain, wherein the parts highlighted by squares belong to the DB$_i$-domain (Fig. 3.24(b)), where recombination is dominated by majority carriers and thus opposite to the standard SRH recombination (more details in Sec. 3.5.2). On a lightly p-type doped wafer surface, without any surface charge Q_s, one has $p_s \geq n_s$, and as a result the recombination rate never lies in the DB$_i$-domain when $\frac{\sigma_n^0}{\sigma_p^0} = \frac{1}{20}$.

3.5. Novel model for a-Si:H/c-Si interface recombination based on the amphoteric nature of silicon dangling bonds

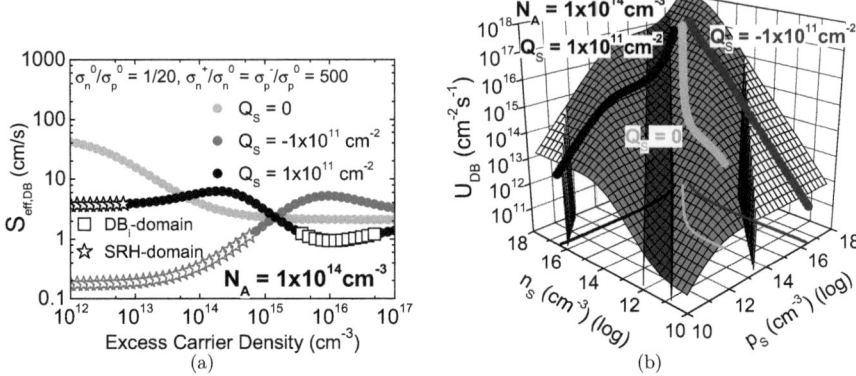

Figure 3.24: *Lightly p-type doped c-Si wafer passivated with a-Si:H inducing varying image surface charges in the c-Si surface $Q_{Si} = -Q_s$. a) Injection level dependent variation of the effective surface recombination velocity calculated from the amphoteric surface recombination model, $S_{eff,DB}$. b) Trajectories on the surface recombination rate's surface plot resulting from surface carrier density couples n_s and p_s, varying accordingly to Q_s. The white surface in b) and the stars in a) denote the SRH-domain, while the dark gray surface in b) and the squares in a) denote the DB_i-domain of the amphoteric surface recombination rate, see text for explanations.*

Conversely, depending on the amount of positive surface charge density, an inversion of the p c-Si surface, i.e. $n_s \geq p_s$, results in a shift of U_{DB} into the DB_i-domain. Finally, negative surface charge density leads to hole accumulation at the p c-Si surface and thus $p_s \gg n_s$ resulting in the recombination rate U_{DB} lying in the SRH-domain.

We conclude that the physical effect of positive resp. negative dangling bond charge in the interface region allows the tuning of the magnitude of field-effect-like passivation. High surface charge densities Q_s, as resulting from a-Si:H doping or external field application, are needed to reach the microscopic picture in which the dangling bond surface recombination rate U_{DB} simplifies to a SRH-like rate. Under our experimental conditions, the amphoteric nature of the DB states dominates recombination and consequently, the standard SRH recombination model does not allow the reproduction of the experimental data presented in Chap. 4.

3.6 Conclusion: comparison of the different interface recombination schemes

Finally, the differences and similarities of the interface recombination schemes are discussed in this section. Similar $\tau_{eff}(\Delta n)$-curves result from different passivation mechanisms, while different interface recombination models can be brought to coincidence while adding a physical meaning to the identified interface parameters.

The absolute lowest interface state density is obtained with thermal SiO_2 passivation schemes [KC02b]. Due to its larger capture cross-section for electrons than for holes, the passivation performance of SiO_2 is superior on n-type c-Si [AGW92]. However, this passivation scheme can encounter long-term stability problems and requires high process temperatures.

The industrially most widespread passivation scheme uses SiN_x passivation, which is dominated by field-effect. Its large inherent positive charge when brought into contact with c-Si leads to low minimal values of the surface recombination velocity when passivating c-Si wafers of all doping types and levels but to inferior device performances than those obtained with SiO_2 [DMMH02]. The performance loss is mainly due to a lower short-circuit current density while the open-circuit voltage is equally high. The predominant loss arises from a parasitic current between the rear metal contacts and the inversion layer induced underneath the SiN_x film by its fixed positive charge inverting the p-type c-Si surface. Electrons in the inversion layer at the rear flow directly into the rear contact instead of being injected into the p-type base of the cell if the voltage across the induced floating junction is small, as is the case in short-circuit conditions. This current path can be regarded as a shunt between the inversion layer and the rear metal contact.

The injection level-dependance of S_{eff} at the SiO_2/c-Si and the SiN_x/c-Si interface is attributed to different passivation properties that are a larger electron than hole capture cross-section at the SiO_2/c-Si interface and a high fixed positive charge in combination with recombination in the space-charge region at the SiN_x/c-Si interface. Nonetheless, the injection level-dependance of S_{eff} at the SiO_2/c-Si and at the SiN_x/c-Si interface are found to be astonishingly similar [SA99]. Biegelsen et al. [BJS+85] already suggested that there are common physical mechanisms underlying the characteristic attributes of the two systems SiO_2/c-Si and bulk a-Si:H justified by their electron spin resonance (ESR) measurements. The amphoteric nature of bulk SiN_x was demonstrated somewhat later also by

3.6. Conclusion: comparison of the different interface recombination schemes

ESR measurements [KLK88]. In the following, a comparison of interface recombination modeling by standard recombination centers and amphoteric interface dangling bonds is conducted.

Yablonovitch et al. [YSEW86] measured the surface recombination at high-quality thermally grown SiO_2/c-Si interfaces as a function of the surface density of electrons and holes, n_s and p_s. For this, semitransparent palladium films were subsequently evaporated on both faces of the wafer to act as gate electrodes. The experiment consisted of measuring the surface recombination velocity S as a function of varying gate bias (-70 V to $+70$ V) at high level injection, where the band bending problem simplifies to the evaluation of a single recombination current instead of the extended SRH formalism (Sec. 3.3.3). Since $S \equiv J(n_s, p_s)/(q\Delta n)$ is not purely a surface property but depends on band bending via the bulk excess carrier density Δn, Yablonovitch et al. introduced the generalized surface recombination velocity $S_{gen} \equiv J(n_s, p_s)/(q\sqrt{n_s p_s})$ [cm/s] which explicitly depends only on surface properties and is independent of band bending. When S_{gen} as a function of $\sqrt{n_s/p_s}$ is modeled with the standard SRH formula (Eq. 3.13), it can simply be written as a sum of simple Lorentzian functions.

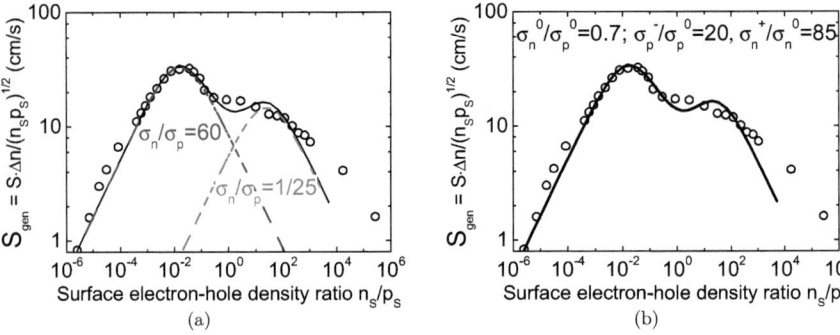

Figure 3.25: *Generalized surface recombination velocity S_{gen} as a function of the surface electron-hole density ratio n_s/p_s at a thermally grown SiO_2/c-Si interface as measured by Yablonovitch et al. [YSEW86] (symbols). Least square fit (solid lines) with a) two standard SRH interface states of two different electron to hole capture cross-section ratios $\frac{\sigma_n}{\sigma_p}$ (dashed lines) and b) amphoteric interface states (Sec. 3.5.3).*

The experimental data of Yablonovitch et al. (Fig. 3.25, symbols)

3.6. Conclusion: comparison of the different interface recombination schemes

may be fitted to least squares (solid curve in Fig. 3.25(a)) by two such Lorentzian functions (dashed curves in Fig. 3.25(a)) with the dominant one peaking at $n_s/p_s = \frac{\sigma_p}{\sigma_n} = 0.017$ and a second one peaking at $n_s/p_s = \frac{\sigma_p}{\sigma_n} = 25$ (in [YSEW86] a different fit with poorer accordance to the measurement points between the 2 peaks is made). The peaks in Fig. 3.25(a) thus represent interface states of two different capture cross-section ratios $\frac{\sigma_n}{\sigma_p}$, the dominant interface states having a capture cross-section ratio $\frac{\sigma_n}{\sigma_p}$ of 60 and the less dominant ones a $\frac{\sigma_n}{\sigma_p}$ ratio of 1/25.

Our modeling of the same data (symbols in Fig. 3.25(b)) with amphoteric interface dangling bonds (solid curve in Fig. 3.25(b)), Sec. 3.5.3, yields exactly the same least square fit as with the standard SRH formula used by Yablonovitch et al.. But a physical meaning can now be attributed to the fit parameters, which show the same trend as that which will be found for a-Si:H/c-Si interface recombination (Chap. 4): a larger neutral hole than electron capture cross-section $\sigma_n^0 < \sigma_p^0$ and much larger charged than neutral capture cross-sections $\sigma^{+,-} \gg \sigma^0$. The larger capture cross-sections of the charged states with respect to the neutral states prevail from a positive correlation energy of the amphoteric interface dangling bonds as measured at the SiO_2/c-Si interface by ESR by Biegelsen et al. [BJS+85]. They found a correlation energy of ~ 0.6 eV slightly larger than the one in bulk a-Si:H. Therefore, the similarity (discussed in Sec. 3.5.2) over a large surface carrier density range between the standard SRH recombination rate with $\sigma_n > \sigma_p$ and the amphoteric recombination rate with $\sigma_n^0 < \sigma_p^0$ and $\sigma^{+,-} \gg \sigma^0$ can be recognized (Fig. 3.20). This similarity is apparent when comparing the capture cross-section ratio value of the dominant peak obtained with the standard SRH formula in Fig. 3.25(a) ($\frac{\sigma_n}{\sigma_p} = 60$) with the fit parameters of the amphoteric recombination model in Fig. 3.25(b) ($\frac{\sigma_n^0}{\sigma_p^0} \times \frac{\sigma_n^+}{\sigma_n^0} = 60$). But obviously, to represent the lower local maximum in the surface plot of the amphoteric recombination rate in Fig. 3.20(c), the lower peak found within the standard SRH modeling (Fig. 3.25(a)) would need to be included in the surface plot corresponding to the standard SRH recombination rate with one single peak in Fig. 3.20(a). Our amphoteric interface recombination model agrees with these previously published data using a simple approach with fit parameters having in our view a more physical meaning.

Krick et al. [KLK88] identified the amphoteric nature of the amorphous silicon-nitride-dangling-bond center in bulk SiN_x by alternate positive and negative charge injection and ultraviolet (UV) illumination. UV illumi-

3.6. Conclusion: comparison of the different interface recombination schemes

nation notably increases the paramagnetic, i.e. neutral, dangling bond density which annihilates any net space-charge previously prevailing in these samples. Under positive (negative) corona charge, electrons (holes) tunnel from the silicon substrate through an underlying thin oxide layer into the nitride where they can be trapped. The loss in ESR signal corresponds to the trapped electron (hole) density that renders these formerly paramagnetic and neutral centers diamagnetic and negatively (positively) charged.

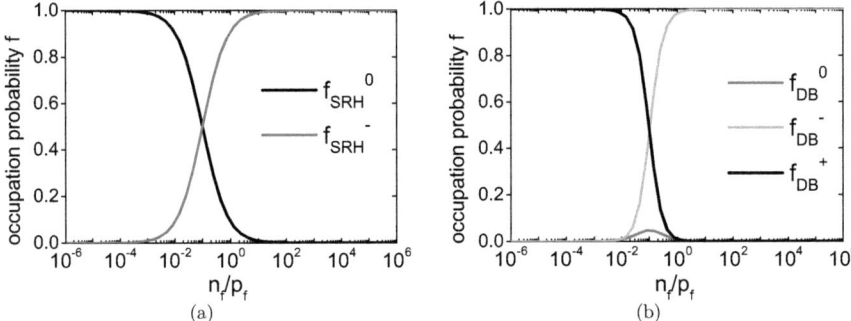

Figure 3.26: *Defect occupation probabilities as a function of the free carrier density ratio n_f/p_f for a) standard SRH recombination with $\frac{\sigma_n}{\sigma_p} = 10$ and b) amphoteric recombination with $\frac{\sigma_n^0}{\sigma_p^0} = 10$ and negative correlation energy E_U resulting for this plot in $\frac{\sigma^{+,-}}{\sigma^0} = 0.1$.*

In as-deposited and programmed nitride memories, there is no ESR signal measurable, and thus, the reaction $2D^0 \rightarrow D^+ + D^-$ is energetically favorable [CLK+90]. Contrary to a-Si:H, most of the amphoteric recombination centers are thus in equilibrium in the charged states. This means that the repulsive energy of having two electrons localized at the same dangling bond site is overcompensated by the energy gain originating from lattice relaxation so that the correlation energy is negative [PR06]. In this case, the standard SRH and the dangling bond occupation probabilities (Eq. 3.41 and Eq. 3.33) shown in Fig. 3.26 become very similar, therefore the surface plots of the corresponding recombination rates shown in Fig. 3.27 and finally $S_{eff}(\Delta n)$-curves measured on passivation samples do become similar too (further explanations are given in Secs. 3.3.1, 3.3.2 and 3.5.2).

3.6. Conclusion: comparison of the different interface recombination schemes

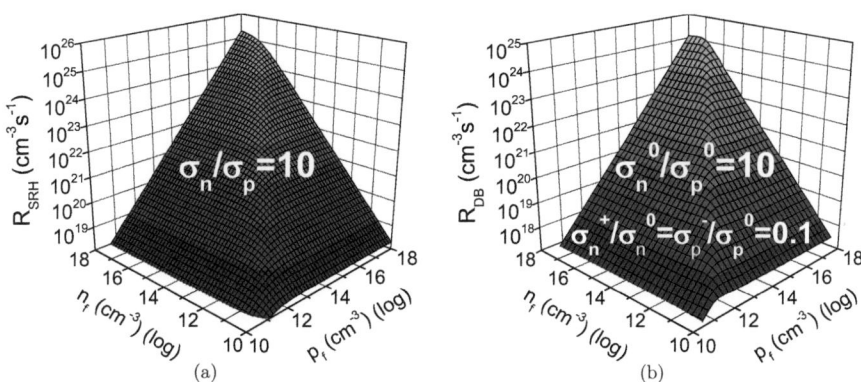

Figure 3.27: *Recombination rates as a function of the free carrier densities n_f and p_f for a) standard SRH recombination with $\frac{\sigma_n}{\sigma_p} = 10$ and b) amphoteric recombination with $\frac{\sigma_n^0}{\sigma_p^0} = 10$ and negative correlation energy E_U for which in this plot the charged to neutral capture cross-section ratios are set to $\frac{\sigma^{+,-}}{\sigma^0} = 0.1$.*

The fixed positive interface charge in SiN_x observed when brought in contact with c-Si then determines the surface carrier densities (n_s, p_s-projections) and finally the trajectories on the recombination rate's surface plot and thus the shape of the $S_{eff}(\Delta n)$-curves measured on passivation samples (c.f. Sec. 3.3.2, e.g. Figs. 3.6 and 3.7 for further explanations).

Also the three different types of defects with broad Gaussian-like density distributions found by Schmidt et al. at the SiN_x/c-Si interface [SA99] can be made to coincide with our DB interface recombination model by considering the appropriate set of recombination parameters for supplementary existing extrinsic DBs. The amphoteric interface recombination model relying on the unique properties of DBs to possess three states of charge, developed in Sec. 3.5.3, is thus potentially applicable to a broader range of heterostructures than only a-Si:H/c-Si to explain interface recombination phenomena within a simple model.

As will be shown in Chap. 4, the growth of intrinsic a-Si:H on c-Si leads to a low interface recombination center density while field-effect passivation can be tuned by varying the average state of charge of the interface's dangling bond recombination centers, for example by an overlying doped thin-film Si layer. Crystalline silicon material of both conduction types

3.6. Conclusion: comparison of the different interface recombination schemes

and all doping levels can thus be effectively passivated by a-Si:H. For example the parasitic shunting at the rear of SiN_x-passivated p-type c-Si solar cells can be circumvented by depositing an i a-Si:H instead of a SiN_x layer for the back surface passivation [DMM+03]. The a-Si:H's suitability for complete c-Si solar cell fabrication is proven by the high-performing HIT (heterojunction with intrinsic thin-layer) cells of Sanyo [TYT+07] that are only outperformed in efficiency (but not in V_{OC}) by Sunpower's back contacted cells based on a high-quality front SiO_2-passivation (capped by SiN) [sun]. When using a-Si:H as a passivation layer for full Si heterojunction solar cell processing on otherwise rather standard c-Si solar cells, the a-Si:H's quality suffers under the high process temperatures potentially used in subsequent c-Si solar cell fabrication process steps. As an alternative, thin films of aluminum oxide (Al_2O_3) [JH85] grown by Atomic Layer Deposition (ALD) rely on excellent field-effect passivation by a high negative charge [AVA+04, ADV+06, HHL+06], in contrast to SiN_x and partially SiO_2 relying on positive field-effect passivation. Thus, the parasitic shunting effect observed at the SiN_x-passivated p-type c-Si solar cell's rear does not occur [SMB+08]. Al_2O_3 is also ideally suited as a front surface passivation layer for n-type c-Si based solar cells neither inducing inversion channel shunting of the p^+-emitter, nor providing parasitic optical absorption in the front surface passivation layer [BHvdS+08]. ALD, however is not suitable for industrial, large-scale production. Its main disadvantage is its low deposition rate which can partially be overcome by depositing ultrathin ALD-Al_2O_3 films and capping them with a thicker film of for example PECVD-SiO_x [SMB+08] or -SiN_x [BHvdS+08].

Finally, from the similarities and various advantages and disadvantages of these different passivation schemes it is becoming obvious that their variety increases the degree in freedom in c-Si solar cell design and that their use in combined stacks is a promising avenue for improved stable passivation schemes in the c-Si wafer industry. As an example, Rohatgi *et al.* [RNR98] first showed a passivation stack consisting of thermal SiO_2/SiN_x. Agostinelli *et al.* [ACD+06] then presented industrially relevant PECVD SiO_x/SiN_x stacks as a rear passivation scheme for standard c-Si solar cells, actually getting introduced in industrial production. These two concepts were later combined to the triple stack layer SiO_x/SiN_x/SiO_x rear passivation scheme [HKS+08]. In another example, a thin a-Si:H layer in stack with a capping SiN_x layer has been shown to give excellent passivation of front diffused emitters of both doping types, the SiN_x layer acting at the same time as an antireflection coating [PTTB06]. As a final example, an a-Si:H/SiO_x stack as rear passivation for standard c-Si solar cells

3.6. Conclusion: comparison of the different interface recombination schemes

shows a rear reflectivity comparable to thermal SiO_2 and appears more stable under different bias light intensities than thermal SiO_2 [HSK+08].

The unique determination of a material's interface passivation properties from the injection level dependent lifetime measurements, i.e. injection level dependent effective surface recombination velocity curves, is difficult because different recombination models and within them different parameter sets can lead to similar calculated $\tau_{eff}(\Delta n)$-curves. For example, recombination at the SiO_2/c-Si and the SiN_x/c-Si interfaces as well as at the intrinsic a-Si:H/c-Si interface (Chap. 4) yields similar measured $\tau_{eff}(\Delta n)$-curves on a variety of c-Si doping levels of both conduction types. It is finally within the complete device, and after the measurement of its electrical performances, that the quality of passivation layers is demonstrated. However, additional characterization such as DLTS, C-V, lifetime measurements on corona-charged samples, HR-TEM allows for faster development of high efficiency passivation schemes for c-Si devices.

Chapter 4
a-Si:H/c-Si interface passivation: experiment & modeling

In the first part of this chapter, experimental lifetime curves of various i a-Si:H passivated c-Si wafers are modeled by the amorphous interface recombination formalism presented in Chap. 3 (Sec. 4.3). Fixing these i a-Si:H passivation layers' outer surface potential by a doped layer effectively adds field-effect passivation (Sec. 4.5). Additionally, the influence of the atomic structure on the a-Si:H/c-Si heterointerface recombination is studied, and the reason for the more challenging passivation of textured c-Si is elucidated by HR-TEM micrographs (Sec. 4.6 and Sec. 4.7). Besides this, the a-Si:H passivation performance on c-Si is compared to the one of SiO_2 and SiN_x (Sec. 4.2), the thickness dependent light soaking behavior of the i a-Si:H passivation is studied (Sec. 4.4), and a roadmap for the choice of the optimal wafer type in view of minimal interface recombination is made (Sec. 4.8). This chapter is supposed to be self-consistent.

4.1 Experiment and modeling

The overall photogenerated carrier lifetime is evaluated via an effective lifetime measurement from which $S_{eff,m}$ can be deduced. In this study the excess carrier density (Δn) dependent effective carrier lifetime $\tau_{eff,m}$ in double-side passivated c-Si is determined with a WCT-100 photoconductance tool from Sinton Consulting [sin] (Sec. 2.1.2.1). In this set-up, the inductively measured excess photoconductance is given by $\Delta \sigma =$

4.1. Experiment and modeling

$q(\Delta n_{av}\mu_n + \Delta p_{av}\mu_p)W$, where $\Delta n_{av} = \Delta p_{av}$ is the average excess carrier density, W the wafer thickness and μ_n, μ_p the well known electron and hole mobilities in c-Si. $\tau_{eff,m}$ is related to Δn_{av} by the time-dependent decay of Δn_{av} and/or the generation rate G_L. The experimentally measured value of the excess carrier density $\Delta n = \Delta n_{av}$ corresponds to the model variable $\Delta n(x=d) = \Delta p(x=d)$ as described in Sec. 3.3.2. The transient photoconductance measurement mode consists of measuring wafer conductivity as a function of time after a very short and intense light pulse. This technique is only appropriate for the evaluation of photogenerated carrier lifetimes appreciably greater than the flash duration. On the contrary, in the quasi-steady-state (QSS) mode, the illumination level dependent wafer conductivity is measured during a long, exponentially decaying light pulse. The so-called generalized analysis of the QSS data allows the characterization of arbitrary lifetimes.

In order to evaluate the effective interface recombination rate over the widest possible range of injection levels, several measurements acquired with both the transient and the quasi-steady-state (QSS) photoconductance techniques [SC96], the latter analyzed in the so-called generalized mode, are combined as shown in Fig. 2.5. This combination of data is the origin of some small discontinuities seen in some of the following measurement data. The transient mode is more appropriate for the measurement of low excess carrier densities and high lifetimes. At higher injection levels, there is good accordance between QSS data, acquired with a long flash and analyzed in the generalized mode and transient data acquired with a short flash. Only in rare cases, larger discrepancies occur at lower injection levels because of the effects mentioned in Sec. 4.3.1 (Fig. 4.21). In such cases no data fitting is performed.

Due to the rather high lifetimes measured and the symmetrical surface passivation schemes used here, the effective surface recombination velocity S_{eff} at the a-Si:H/c-Si interface is related to τ_{eff} by the simplified equation (Sec. 3.1):

$$S_{eff} = \left(\frac{1}{\tau_{eff}} - \frac{1}{\tau_{bulk}}\right) \cdot \frac{W}{2}, \qquad (4.1)$$

where τ_{bulk} is the bulk c-Si lifetime (Sec. 3.2) composed of the extrinsic and intrinsic wafer lifetimes $\frac{1}{\tau_{bulk}} = \frac{1}{\tau_{extr}} + \frac{1}{\tau_{intr}}$. We set the extrinsic, i.e. the defect limited lifetime according to the ingot lifetime specified by the wafer supplier or otherwise to one of the highest ever measured lifetime value of 37 ms by Yablonovitch et al. [YAC+86] (Sec. 3.2). τ_{intr} governed by Auger and radiative recombination is given by Eq. 3.10 in Sec. 3.2.

4.1. Experiment and modeling

In order to be able to extract criteria for the quality and the nature of interface passivation, i.e. the values of the interface state density N_s and the interface charge density Q_s, we fit the measured $S_{eff,m}(\Delta n)$ with our model described by $S_{eff,DB,c} = f(\Delta n; n_0, p_0; Q_s; N_s; \sigma_p^0, \frac{\sigma_n^0}{\sigma_p^0}, \frac{\sigma^{+,-}}{\sigma^0})$, i.e. Eq. 3.44 with Eqs. 3.24, 3.22 and 3.20. For this purpose, the values of the capture cross-section ratios $\frac{\sigma_n^0}{\sigma_p^0}$, $\frac{\sigma_n^+}{\sigma_n^0}$ and $\frac{\sigma_p^-}{\sigma_p^0}$ have to be known. If one refers to values published in the literature [HSSP95], their range is so wide that it does not permit a fit of experimental data with our model. Thus, here we measure the injection level-dependance of the effective surface recombination velocity and choose capture cross-section ratios which allow for a reasonable fit of all our experimental data simultaneously. Best accordance is obtained by trial and error, trials in which we "manually" introduce various sets of published values for $\frac{\sigma_n^0}{\sigma_p^0}$ and $\frac{\sigma_n^+}{\sigma_n^0} = \frac{\sigma_p^-}{\sigma_p^0}$ [HSSP95]. The best fits of the experimental data of i a-Si:H layers on various p- and n-type wafers with our model, yield a neutral electron to hole capture cross-section ratio of $\frac{\sigma_n^0}{\sigma_p^0} = \frac{1}{20}$ and charged to neutral capture cross-section ratios of $\frac{\sigma_n^+}{\sigma_n^0} = \frac{\sigma_p^-}{\sigma_p^0} = 500$. The corresponding capture cross-section hierarchy $\sigma_n^0 < \sigma_p^0 < \sigma_n^+ < \sigma_p^-$ (1 : 20 : 500 : 100'000) is the same as the one found by Street [Str84] (although less pronounced, in his case 1 : 3 : 5 : 7) when considering the influence of disorder and band-tail localization on deep trapping processes in amorphous semiconductors. A similar neutral electron to hole capture cross-section ratio of $\frac{\sigma_n^0}{\sigma_p^0} = \frac{1}{4}$ is also obtained by using the experimentally observed ratio of the a-Si:H's mobility-lifetime $\mu\tau^0$ product $\frac{\mu_n \tau_n^0}{\mu_p \tau_p^0} = 10$ [SZT83], considering a value of $\frac{\mu_n}{\mu_p} = 3$ for the ratio of the mobilities in a-Si:H [TDRiCC04] and the slight asymmetry in the thermal velocity ratio of electrons to holes $\frac{v_{thn}}{v_{thp}} = 1.22$.

Once the microscopic parameters are fixed to their pre-evaluated values, the two remaining model parameters N_s and Q_s, giving the best accordance between the model and the experimental data for each case, are again obtained by "manual" adjustment. Such a procedure is not time consuming here, as (Sec. 3.3.2) the magnitude of N_s only scales the whole curve by a constant factor, while Q_s gives the shape of the $S_{eff,c}(\Delta n)$-curve. To summarize, MATLAB calculation input parameters are the capture cross-sections σ_n^0, σ_p^0, σ_n^+, σ_p^- and the thermal velocity v_{th}, and in the following we search the set of a-Si:H/c-Si interface parameters N_s and Q_s to find those which give a reasonable agreement between the calculated $S_{eff}(\Delta n)$-curve and the measured one. Thus we expect to find the microscopic mechanism limiting the interface recombination.

First, we investigated the passivation performance of intrinsic a-Si:H on c-Si with an a-Si:H layer thickness of 5 nm as used in heterojunction solar cell devices [FK05]. However, such thin i a-Si:H layers are never used without capping layers. Therefore, in order to evaluate the passivation performance of i a-Si:H, much thicker (40 nm thick) layers were grown on various c-Si doping type and level wafers. These layers are about twice as thick as the typical total thickness of stacks grown for our heterojunction solar cell fabrication. The a-Si:H thickness series are studied on two different wafers. To investigate the influence of the well known a-Si:H bulk degradation under light soaking (Staebler-Wronski effect [SW77]) on the a-Si:H/c-Si interface passivation, these series are used for light and dark degradation studies. Further on, we intentionally change the average dangling bond charge state of 40 nm thick i a-Si:H layers by microdoping, thus varying Q_s without increasing N_s. The same additional field-effect passivation is achieved by growing stacks of i a-Si:H plus doped layers. HR-TEM micrographs allow the observation of atomic interfaces and the material system's structure. Finally, the influence of surface texture on the a-Si:H passivation quality, moreover of the texture's specific morphology, is studied.

In the following, the measured $S_{eff,m}(\Delta n)$-curves are indicated by symbols, while the $S_{eff,DB,c}(\Delta n)$-curves, calculated with the model parameter couples N_s, Q_s found to give best accordance between measured and calculated injection level dependent surface recombination, are indicated by lines. Alternatively, when $\tau_{eff,m}$ approaches τ_{bulk}, one sees from Eq. 4.1 that the determination of $S_{eff,m}$ becomes affected by large incertitudes (c.f. Fig. 3.9 and related discussions). That is why mostly we prefer to plot $\tau_{eff}(\Delta n)$-curves, including the theoretical limit set by τ_{bulk} (dashed lines), and we do not consider the high injection level range for curve fitting.

4.2 State of the art Si surface passivation

The crystalline Si bulk lifetime is maximal for lightly doped (intrinsic recombination is then only limiting lifetimes at high injection levels) float zone (FZ, few bulk defects, i.e. low extrinsic recombination) wafers. As these show thus maximal sensitivity to surface recombination, they are the most appropriate for exploring the upper lifetime limits imposed by surface recombination of different passivation schemes. Figure 4.1 compares the passivation performance of our intrinsic a-Si:H with those of SiO_2 and SiN_x

4.2. State of the art Si surface passivation

grown by Kerr et al. [KC02b, KC02a]. The intrinsic c-Si bulk lifetime is parameterized according to Kerr et al. [KC] and the extrinsic c-Si bulk lifetime's upper limit is set according to Yablonovitch et al. [YAC+86] to τ_{extr} = 37 ms (Sec. 3.2). Alnealed thermally grown SiO_2 layers yield highest measured lifetimes. SiN layers (stoichiometric) achieve slightly lower passivation performances. Our VHF-PECVD intrinsic a-Si:H passivation has almost the same quality as Kerr's SiN passivation. All three passivation schemes reach the same maximal lifetimes on lightly p- and lightly n-type doped c-Si.

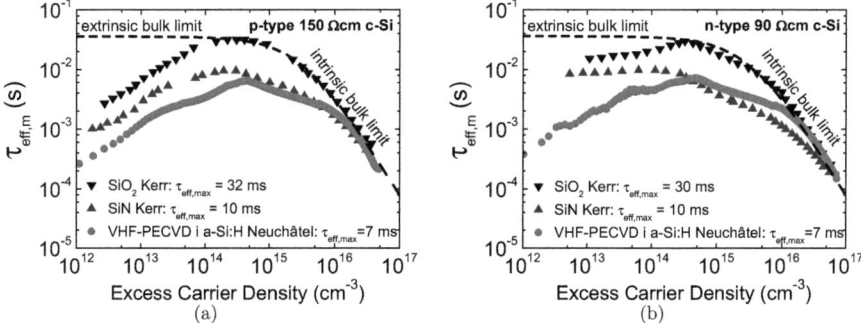

Figure 4.1: *State of the art surface passivation of lightly doped a) p- and b) n-type FZ c-Si with SiO_2, SiN and a-Si:H.*

It is worth mentioning here that our wafers undergo no special cleaning step before VHF-PECVD deposition, they are only dipped in diluted HF. Note also that the calibration of the WCT-100 photoconductance instrument is made at low conductance [MGAB08]. If the calibration factor depends markedly on wafers conductance, this can lead to measurement errors in the injection level-dependance of $\tau_{eff,m}$ as will be discussed in Sec. 4.3.1. But the comparison remains nonetheless valid because all curves are measured in the same configuration by the Sinton lifetime tester and high injection level lifetimes are correctly measured (Fig. 4.9 and related comments).

4.3 Intrinsic a-Si:H on various flat c-Si substrates

The high passivation performance of i a-Si:H on all kinds of flat c-Si substrates is shown in this section. The validity of our amphoteric interface recombination model is demonstrated by the analysis of the data presented here. In addition, experimental lifetime measurement issues and the limits of our interface recombination model are addressed.

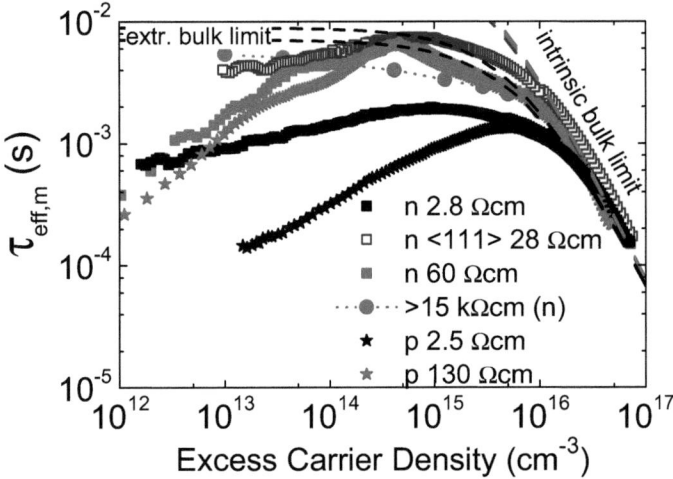

Figure 4.2: *Effective lifetime measurement of 45 nm thick i a-Si:H layers on flat c-Si of various doping types and levels. Dashed lines show the lifetime limit imposed by c-Si bulk recombination for the corresponding wafers (Sec. 3.2).*

Fig. 4.2 shows the measured injection level dependent lifetimes of 45 nm thick i a-Si:H layers symmetrically passivating flat c-Si wafers of varying doping types and levels. All $\tau_{eff,m}(\Delta n)$-curves are measured with the Sinton lifetime tester except the one on the almost intrinsic c-Si wafer, which is evaluated from ILM mappings (Sec. 2.1.2.1) at different illumination levels (and therein the lowest injection level point (diamond in Fig. 4.7) originates from MW-PCD-measurements).

When representing $S_{eff,m}(\Delta n)$-curves we can make abstraction from the wafer thickness-dependance that causes a "visual overestimation" when

4.3. Intrinsic a-Si:H on various flat c-Si substrates

looking only at $\tau_{eff,m}(\Delta n)$-curves of the i a-Si:H surface passivation quality for the 130 Ωcm p-type wafer having a thickness of 525 μm, twice that of the other wafers having thicknesses between 250 and 300 μm (Fig. 4.3). For example, $\tau_{eff,m,max}$ on lightly p- and n-type doped c-Si that reaches experimentally the same 7 ms value, translates via Eq. 4.1 into a $S_{eff,m,min}$ of 3 cm/s on the thicker lightly p-type vs 1.5 cm/s on the thinner lightly n-type doped wafer (compare corresponding curves in Figs. 4.2 and 4.3) at an injection level of $\Delta n = 5 \times 10^{14}$ cm^{-3} where bulk c-Si recombination is negligible on both wafers. For high measured lifetimes, we choose to display τ_{eff}-curves because of the incertitudes introduced by the choice of the intrinsic bulk lifetime parametrization in the evaluation of $S_{eff,m}$ at high injection levels (Fig. 3.9 and related discussion).

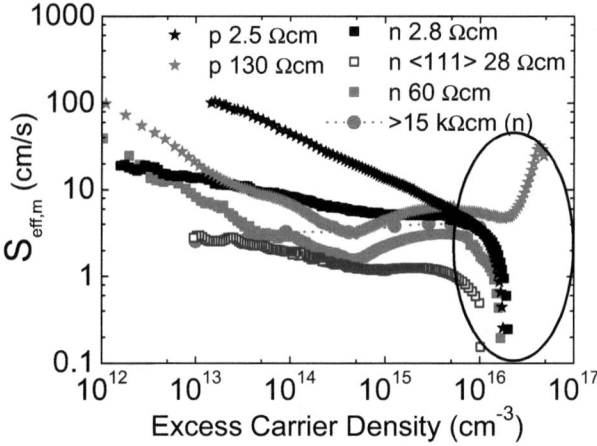

Figure 4.3: *Passivation performance of 45 nm thick i a-Si:H layers on various flat c-Si wafers, calculated from the effective lifetime measurement shown in Fig. 4.2 and the parametrization of bulk c-Si recombination described in Sec. 3.2. For such high measured lifetimes, the choice of bulk lifetime parametrization dominates the values of $S_{eff,m}$ at high injection levels.*

Figure 4.4 represents the same lifetime measurement data for i a-Si:H passivating various c-Si wafers as in Fig. 4.2 (symbols) including fits to the experimental curves obtained with our amphoteric interface recombination model (lines). The extracted values of the interface dangling bond density N_s and the charge density Q_s for the different wafers are listed in Tab. 4.1.

4.3. Intrinsic a-Si:H on various flat c-Si substrates

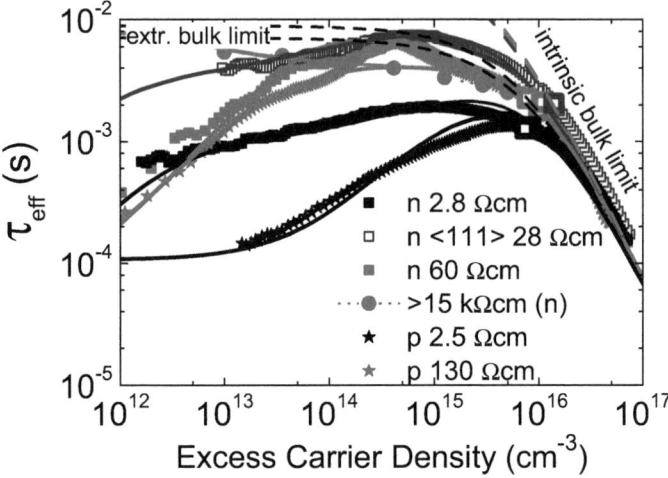

Figure 4.4: *Lines: fits with our amphoteric interface recombination model (Sec. 3.5.3) to the experimental $\tau_{eff,m}(\Delta n)$-curves from Fig. 4.2. Data measured for 45 nm thick i a-Si:H passivating differently doped flat c-Si. Model parameters N_s and Q_s giving the best agreement between theory and experiment are listed in Tab. 4.1. Big squares indicate the $\Delta n, \tau_{eff,m}$-couples corresponding to 1-sun illumination, which permit the calculation of the upper limits imposed on a completed solar cell's V_{OC} by recombination (Tab. 4.2).*

45nm i a-Si:H on	N_s $[10^9 \text{ cm}^{-2}]$	Q_s $[10^{10} \text{ cm}^{-2}]$
n 2.8 Ωcm	1.0	-2.2
n <111> 28 Ωcm	0.45	-0.5
n 60 Ωcm	1.6	+1.1
> 15 kΩcm (n)	1.6	+0.1
p 2.5 Ωcm	1.4	+1.8
p 130 Ωcm	3	+0.5

Table 4.1: *Model parameter couples N_s, Q_s giving best agreement between measured and calculated injection level dependent surface recombination for 45 nm thick i a-Si:H layers passivating various flat c-Si wafers. $\frac{\sigma_n^0}{\sigma_p^0} = \frac{1}{20}$ and $\frac{\sigma_n^+}{\sigma_n^0} = \frac{\sigma_p^-}{\sigma_p^0} = 500$ are set constant for all fits.*

4.3. Intrinsic a-Si:H on various flat c-Si substrates

The fit accuracies and the values extracted for N_s in the mainly considered medium illumination level range depend on the values assumed for extrinsic bulk recombination that is set to one of the highest ever measured lifetime value of 37 ms of Yablonovitch et al. [YAC+86] (Sec. 3.2). The values of N_s given in Tab. 4.1 are thus upper limits of N_s. If we assume for the 2.5 Ωcm p-type c-Si wafer, an extrinsic bulk lifetime of 3.4 ms instead of 37 ms, which is in fact more realistic, we can obtain a better fit with N_s reduced from 1.4×10^9 cm^{-2} to 1.1×10^9 cm^{-2}, except for the remaining high injection level discrepancies prevailing from incertitudes in the intrinsic c-Si bulk lifetime calculation, as shown in Fig. 4.5.

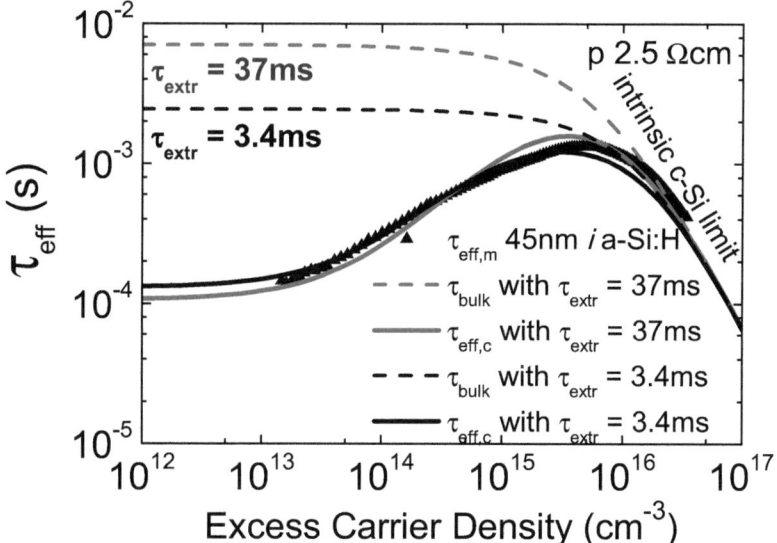

Figure 4.5: *Illustration of the effect of the choice of the extrinsic lifetime τ_{extr} on fitting to experimental $\tau_{eff,m}(\Delta n)$-curves for the case of 45 nm thick i a-Si:H passivating 2.5 Ωcm p-type c-Si.*

However, the value for N_S is not too sensitive ($\pm 30\%$) to the value of the assumed bulk lifetime in this high bulk lifetime value range. From Tab. 4.1 it follows that the fit values found for Q_s are determined by the wafer doping type and level. This finding confirms the amphoteric nature of c-Si surface dangling bonds. The influence of the c-Si's doping type and level on the surface passivation mechanism of i a-Si:H can be explained for the three curves corresponding to the highest doped c-Si wafers (without

4.3. Intrinsic a-Si:H on various flat c-Si substrates

calibration artifact): passivation of the 28 Ωcm, thus rather lightly doped n-type wafer with i a-Si:H results in the diffusion of electrons (majority carriers) into the a-Si:H where they occupy bandgap states (band diagram in Fig. 4.6(a)). Therefore, the average state of charge of the DBs in the vicinity of the i a-Si:H interface has become more negative and a positively charged space-charge region has been left at the c-Si surface.

From the more strongly n-type doped wafer (2.8 Ωcm) the electron diffusion current into the i a-Si:H is larger. Therefore, the average state of charge of the DBs in the i a-Si:H near the interface has become more negative and the charge density in the space-charge region has become more positively charged, Fig. 4.6(b). If the c-Si wafer is p-type doped, e.g. with 2.5 Ωcm resistivity in this case, holes diffuse into the i a-Si:H, leading to a positive average DB charge state and leaving behind in the c-Si a negatively charged space-charge region (Fig. 4.6(c)). Besides the wafer doping, the amount of the average i a-Si:H DB charge state depends on the band offset between a-Si:H and c-Si, set in Fig. 4.6 in accordance with most literature [SKL+07, SDGC+95] to a conduction band offset ΔE_C of 0.2 eV resulting in a valence band offset ΔE_V of 0.4 eV when assuming an i a-Si:H bandgap value of 1.7 eV, see Fig. 5.4. The lower absolute amount of calculated interface DB charge, i.e. $Q_{s,p\,c-Si} = +1.8 \times 10^{10}$ cm^{-2} vs $Q_{s,n\,c-Si} = -2.2 \times 10^{10}$ cm^{-2} on the more highly doped p-type (2.5 Ωcm ∼ $N_A = 6 \times 10^{15}$ cm^{-3}) than n-type c-Si (2.8 Ωcm ∼ $N_D = 2 \times 10^{15}$ cm^{-3}), supports the hypothesis of a larger valence band offset for this i a-Si:H layer passivating c-Si.

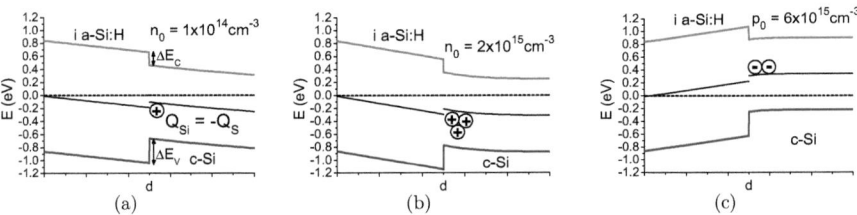

Figure 4.6: *Band diagram of i a-Si:H passivating a) lightly doped n-type c-Si (28 Ωcm), b) more highly doped n-type c-Si (2.8 Ωcm) and c) rather highly doped p-type c-Si (2.5 Ωcm). The corresponding relative amount of charge densities induced in the c-Si surface ($Q_{Si} = -Q_s$) are in good accordance with the values of the surface charge densities Q_s listed in Tab. 4.1, which give the best accordance between theoretical and experimental $\tau_{eff}(\Delta n)$-curves.*

4.3. Intrinsic a-Si:H on various flat c-Si substrates

Photoconductance in the stacked structure and solar cell open-circuit voltage are both ultimately measurements of the same excess carrier density. Thus, an implied open-circuit voltage (implV_{OC}) value can be calculated from the lifetime data [SC96] (Sec. 2.1.2.1, Eq. 2.13). Its 1-sun value allows a useful determination of the upper limits imposed on V_{OC} by recombination and is thus of direct use in further heterojunction solar cell processing. The $\Delta n, \tau_{eff,m}$-couples corresponding to 1-sun illumination are marked in Fig. 4.4 and Tab. 4.2 summarizes the implied V_{OC}s. Open-circuit voltages higher than 700 mV are implied on all these c-Si substrates. Some of them are even no longer limited by interface recombination but by intrinsic c-Si recombination (Sec. 3.2), and are thus at the theoretical limit.

45nm i a-Si:H on	implV_{OC} [mV]
n 2.8 Ωcm	715
n <111> 28 Ωcm	735
n 60 Ωcm	725
> 15 kΩcm (n)	715
p 2.5 Ωcm	715
p 130 Ωcm	705

Table 4.2: *ImplV_{OC} calculated from the measured 1-sun lifetimes of various flat c-Si substrates passivated by 45 nm thick i a-Si:H layers ($\Delta n, \tau_{eff,m}$-couples indicated in Fig. 4.4).*

The interface dangling bond densities N_s are similarly low on all inspected c-Si wafers, in which the single <111> crystal orientated c-Si wafer shows an effective surface recombination velocity of 1 cm/s (corresponding to a highest lifetime of 7.5 ms) and therefore the lowest surface dangling bond density when passivated with i a-Si:H. Interestingly, also HF-passivation performs best on <111> orientated surfaces [YG86], whereas for SiO_2 and SiN_x the passivation of <100> orientated c-Si is superior [RD79, SSES96]. In view of the targeted application of passivation in textured monocrystalline Si solar cells that present <111>-oriented pyramidal surfaces, this is an important observation.

4.3. Intrinsic a-Si:H on various flat c-Si substrates

4.3.1 Hardware and physical effects affecting the measurements

In some measurement configurations erroneous data can be obtained if no care is taken.

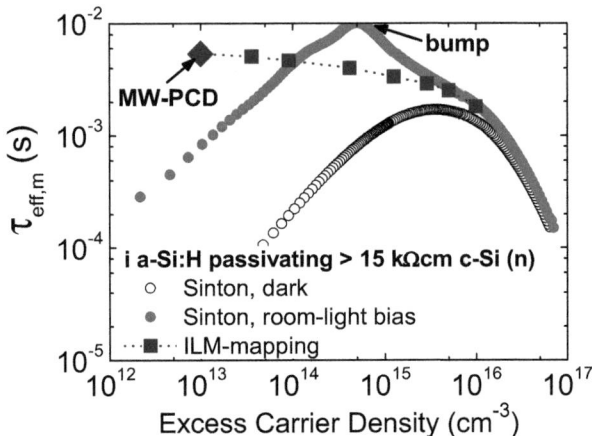

Figure 4.7: *i a-Si:H passivated intrinsic c-Si ($n > 15$ kΩcm): Sinton lifetime measurements made in the dark (open dots) and under room light illumination (full dots) compared to ILM (squares) measurements on the same sample (Fig. 4.8), and MW-PCD, the diamond. The discrepancies result from the calibration of the WCT-100 photoconductance tool. Thus, the ILM measurement is adopted for the comparison of the passivation performance of i a-Si:H on different wafers in Fig. 4.2.*

The dots in Fig. 4.7 show the lifetime decrease at lower injection levels that is artificially measured by the Sinton lifetime tester on such an almost intrinsic c-Si wafer. Transient and generalized QSS measurements and measurements made with different initial flash light intensity attenuation-filters overlap perfectly when measured with the Sinton lifetime tester. But the onset of the lifetime decrease towards low injection levels depends strongly on the bias light intensity. Figure 4.7 illustrates measurements made with room light (full dots) and without (almost in the dark, open dots). The discrepancies between the two curves result from the calibration of the WCT-100 photoconductance tool: the Sinton lifetime tester converts the coil's response into a voltage signal and via a calibration factor, into a sheet conductance. To calculate the calibration factor, a linear relationship be-

4.3. Intrinsic a-Si:H on various flat c-Si substrates

tween the output voltage and the conduction is presumed, but for sample conductances under 1 mS, this relationship becomes non-linear [MGAB08]. With the application of a light bias, the output voltage vs sample conductivity is longer in the linear region of the calibration curve. During the span of a Sinton lifetime measurement, and only for lower injection levels, it drops back in the non-linear region, producing lifetime evaluation errors (compare full to open dot curve in Fig. 4.7).

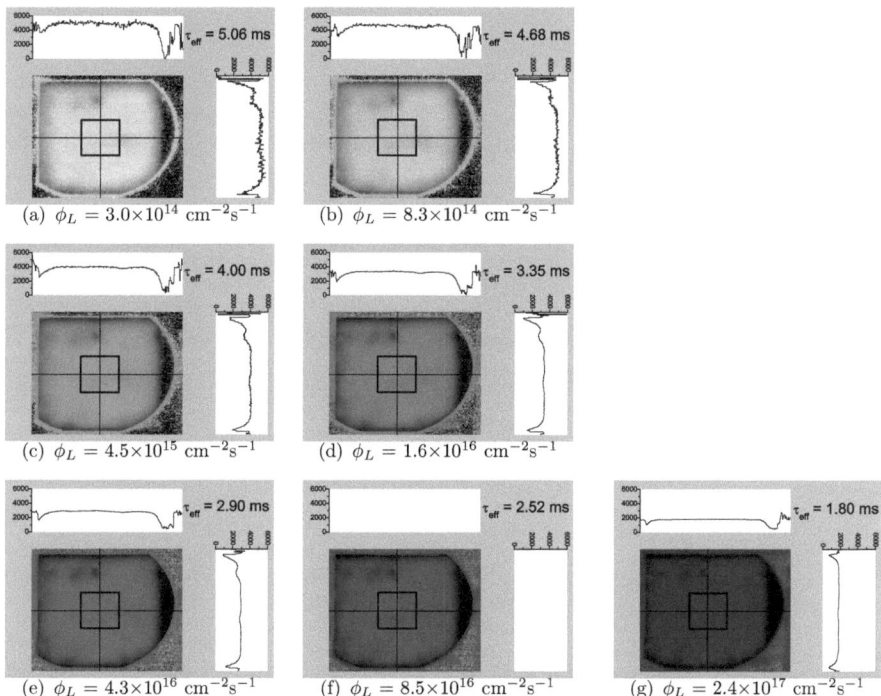

Figure 4.8: *ILM measurements made under varying photon flux densities ϕ_L, increasing from a) to h). The square with 3 cm side length denotes the region over which $\tau_{eff,m}$ is averaged to calculate the excess carrier density Δn from a given photon flux density (Eqs. 2.9 and 2.10), yielding the $\tau_{eff,ILM}(\Delta n)$-curve in Fig. 4.7.*

To unambiguously determine the injection level dependent lifetime, ILM mappings under a varying photon flux density were performed as shown in Fig. 4.8. To calculate the excess carrier density from a given

4.3. Intrinsic a-Si:H on various flat c-Si substrates

photon flux density, $\tau_{eff,m}$ is averaged over a 3×3 cm^2 square corresponding about to the surface measured by the Sinton lifetime tester (indicated by a black square in Fig. 4.8). From Eqs. 2.9 and 2.10 the corresponding excess carrier density is then calculated yielding the $\tau_{eff,ILM}(\Delta n)$-curve consisting of the eight squares in Fig. 4.7. The lowest injection level value of this curve is measured by MW-PCD and highlighted by a diamond.

In addition, there is a bump in the lifetime curves measured on the lightly p- and n-type doped wafers (Fig. 4.2 or more visible on Fig. 4.7). This bump is believed to be again due to the low dark conductance of these samples, corresponding to the non-linear regime of the calibration curve of the Sinton lifetime tester. The application of a corrected calibration curve can overcome this problem, but is not trivial [MGAB08]. Besides the above mentioned application of a bias light to shift the measured curve further into the linear region, the deposition of an ultrathin semitransparent conducting layer on top of the i a-Si:H layer causes the overall sample conductance to lie in the linear region of the calibration curve as shown in Fig. 4.9.

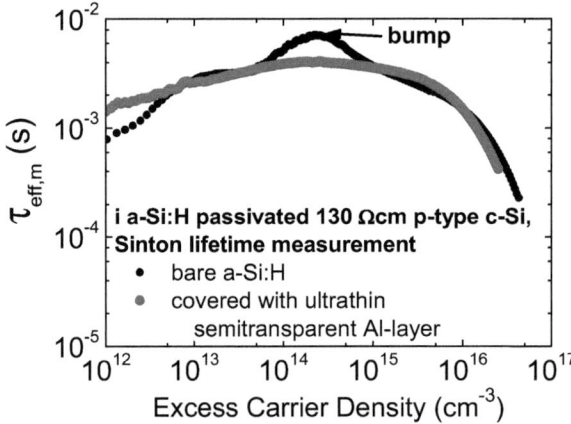

Figure 4.9: *Sinton lifetime measurement made on i a-Si:H passivated lightly p-type doped c-Si with an ultrathin semitransparent conducting layer on top (gray dots). Such a highly conductive capping layer shifts the measured lifetime curve into the linear region of the calibration curve and thus eliminates the bump in the lifetime curves measured on lightly doped c-Si (black dots).*

4.3. Intrinsic a-Si:H on various flat c-Si substrates

However, the deposition of such a capping metal layer can induce a-Si:H/c-Si interface defects and/or introduce field-effect related passivation changes due to the work function difference at the metal/i a-Si:H interface.

The condition of a medium illumination level used to obtain Eq. 3.37 is no longer valid at low injection levels. The reason for this is that at low injection levels the energy interval limited by the quasi-Fermi levels for traps $[E_{tn},E_{tp}]$ within which electronic states act as recombination centers (and not as traps), no longer includes the whole DB distribution (Fig. 3.15). Figure 4.10 shows a sketch of the injection level dependent position of the quasi-Fermi level for traps E_{tn} and E_{tp}, as calculated from Eq. 3.29 and Eq. 3.30 for the lightly p-type doped wafer (with the fit values given in Tab. 4.1).

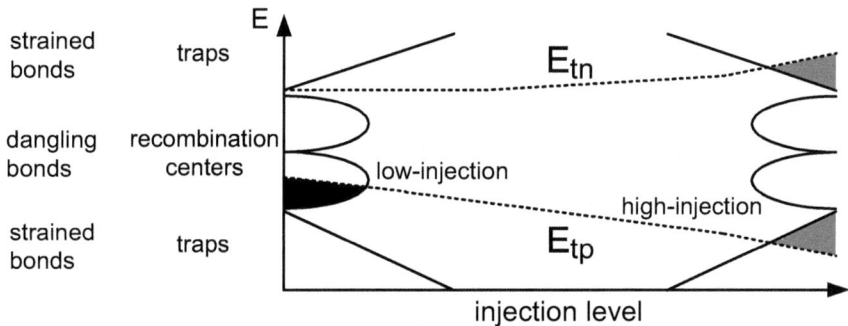

Figure 4.10: *Injection level dependent bandgap position of the quasi-Fermi levels for electron and hole traps, E_{tn} and E_{tp}. To find the closed-form dangling bond recombination rate (Eq. 3.37), the interval $[E_{tn},E_{tp}]$ must properly separate traps from recombination centers. At low and high injection levels this condition is no longer fulfilled, shown schematically by the filled-in surfaces.*

Thus, when fitting experimental data to our dangling bond interface recombination model, one has to pay attention to the validity range of the used amphoteric recombination rate (Sec. 3.5.2). At low Δn (i.e. low injection levels), the number of DBs acting as recombination centers is decreasing (filled-in black area in Fig. 4.10). The calculated interface recombination is thus overestimated in low injection level conditions, because $S_{eff,DB,c}$ given by Eqs. 3.44, 3.20, 3.24 and 3.22 is simply proportional to N_s. Figure 4.10 simultaneously shows that E_{tn} and E_{tp} also do not properly separate traps from recombination centers in high injection level conditions: E_{tn} and E_{tp} enter into the bandtails and cause trap states to act as recombination centers at high injection levels (filled-in gray area in Fig. 4.10 at high injection levels). Therefore, the calculated interface recombination is underestimated in high injection level conditions, contrariwise to low injection level conditions. In practice, at low injection levels, the measured lifetime data becomes noisy and at high injection level densities the choice of the Auger parametrization to calculate the bulk c-Si lifetime dominates the calculated lifetime (Fig. 3.9). Thus, we seek the best accordance between experiment and modeling in the medium illumination level range, which is also the most relevant for our work.

4.4 Intrinsic a-Si:H of varying thicknesses

In this section, the i a-Si:H/c-Si interface passivation performance's dependance on the i a-Si:H layer thickness is examined. Being aware of the light degradation issue of a-Si:H, the effect of light-soaking on interface recombination is studied. The sensibility of our uncovered ultra-thin i a-Si:H layers to ambient atmosphere will be identified and discussed further.

4.4.1 a-Si:H thickness dependent passivation

In c-Si solar cells mainly ultrathin 4 to 10 nm thick i a-Si:H passivation layers are used. On one hand, a minimum thickness is needed to ensure a good passivation, on the other hand light absorption in the i a-Si:H has to be minimized and the carrier transport has to be ensured [TMT08].

Figure 4.11 shows the thickness-dependance of the passivating properties of a-Si:H i-layers deposited on both sides of 60 Ωcm n- and 130 Ωcm p-type c-Si substrates. $\tau_{eff,m}(\Delta n)$-curves are measured directly after annealing of the samples (standard anneal requiring 90 min at 180 °C in a nitrogen atmosphere) and the corresponding $S_{eff,m}$ values at an injection level of $\Delta n = \Delta p = 1 \times 10^{15}$ cm^{-3} are displayed in Fig. 4.11. This passivation series used for thickness dependent light and dark degradation study (Secs. 4.4.2 and 4.4.3) is issued from the initial passivation tests, when optimized layers and fabrication procedures were not established yet. Therefore, measured lifetimes are lower than the best reported data (c.f. Sec. 4.3). By consequence, bulk c-Si recombination has a minor influence on the value of $S_{eff,m}$ even at high injection levels (Eq. 4.1). In addition, the thicknesses of the two wafers used for this study differ by a factor two

4.4. Intrinsic a-Si:H of varying thicknesses

and thus do so $\tau_{eff,m}$ values corresponding to the same $S_{eff,m}$ value (again Eq. 4.1, see also the numerical example in Sec. 3.4.1). Hence the representation in terms of $S_{eff}(\Delta n)$- instead of $\tau_{eff}(\Delta n)$-curves is chosen in this section.

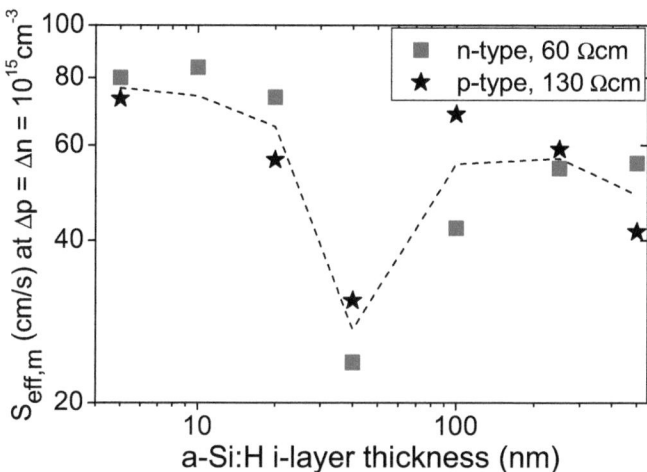

Figure 4.11: *a-Si:H i-layer thickness dependent surface passivation of lightly n- and p-type doped c-Si substrates. $S_{eff,m}$ is evaluated directly after annealing by the generalized QSSPC technique at $\Delta p = \Delta n = 1 \times 10^{15}$ cm^{-3}. The dashed line is a visual guide.*

Similar passivation performances, i.e. similar $S_{eff,m}$ values at an injection level of $\Delta p = \Delta n = 1 \times 10^{15}$ cm^{-3}, are achieved on double-side passivated wafers of both lightly doped n- and p-type. The a-Si:H i-layer thickness has a major effect on the observed $S_{eff,m}$ value, with a minimal value for an i-layer thickness around 40 nm. $S_{eff,m}$ is reduced by increasing the thickness of the passivation layer from 5 nm to 40 nm. This observation is explained with a technological phenomena, the production of surface defects by the initial interface between the plasma and the c-Si surface. Subsequent growth of the a-Si:H layer leads, due to structural relaxation, to a reduction of these plasma-induced defects at the c-Si surface. This relaxation effect still occurs for a-Si:H layers that are about 30 nm thick [NHK93]. For i a-Si:H layer thicknesses increasing from 40 nm to 500 nm, we observe again an increase of $S_{eff,m}$. The increasingly more defective interface may be related with increasing mechanical stress at the c-Si/a-Si:H interface when growing such thick layers [WK88, Cha96]. Indeed, it is

4.4. Intrinsic a-Si:H of varying thicknesses

observed that 1 μm thick a-Si:H layers can peel off when deposited on c-Si wafers.

Surface recombination comparisons at a single fixed injection level are not valid in general, as e.g. in Fig. 4.25 $\tau_{eff,m}(\Delta n)$-curves cross and thus the choice of the best passivation layer depends on the injection level one looks at. However, within this series the most drastic $S_{eff,m}(\Delta n)$-curve shape change shown in Fig. 4.12(a) is still so that the curves do not cross, and thus here, a comparison of $S_{eff,m}$ at $\Delta n = 1 \times 10^{15}$ cm^{-3} at least shows up the right tendencies. For example after 10 hours of dark storage in ambient atmosphere (Sec. 4.4.3) the $S_{eff,m}(\Delta n)$-curves for three different i-layer thicknesses shown in Fig. 4.12(b) are just shifted by a constant factor and can therefore be modeled by a change solely in the interface DB density.

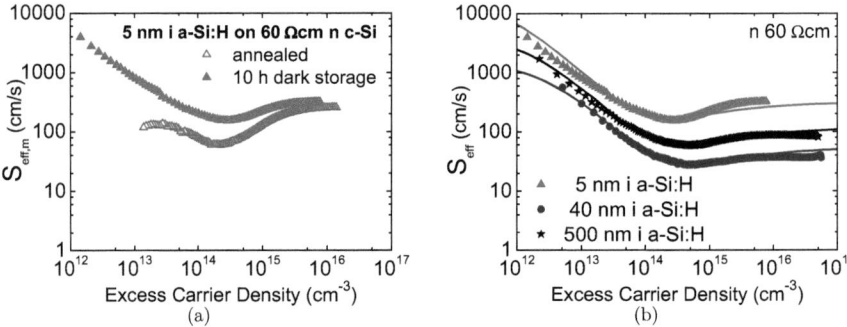

Figure 4.12: *Injection level dependent passivation properties of a-Si:H i-layers on lightly n-type (60 Ωcm) c-Si a) for the thinnest a-Si:H i-layer after annealing and dark storage and b) for varying a-Si:H i-layer thicknesses after 10 hours of dark storage.*

4.4.2 Light degradation

The effect of light soaking on a-Si:H is well known: carrier recombination leads to an increased recombination center density (dangling bonds) and consequently to a degradation of the material's electronic quality [SW77]. The resulting performance degradation of amorphous silicon solar cells depends on the thickness of the intrinsic a-Si:H layer, that is, on the total free carrier recombination rate that increases with the i-layer thickness. pin devices with thin i-layers are thus more stable because of their more

4.4. Intrinsic a-Si:H of varying thicknesses

efficient carrier extraction. Using the thickness series of a-Si:H i-layers ranging from 5 to 500 nm, deposited on both surfaces of lightly doped n- and p-type FZ c-Si substrates, the effect of light soaking on the passivation properties can be quantified. Light soaking is performed under AM1.5g spectra at a temperature of 50 °C in ambient atmosphere (typical a-Si:H light soaking conditions). Figure 4.13 shows the effect of light soaking on the normalized surface recombination velocity (with respect to the annealed value) $S_{eff,m}/S_{eff,ann}$ at $\Delta p = \Delta n = 1 \times 10^{15}$ cm^{-3} for a selection of a-Si:H i-layer thicknesses of 5 nm, 40 nm and 500 nm on n-type c-Si (60 Ωcm). The same trend is observed for i a-Si:H passivated p-type c-Si (130 Ωcm) substrates.

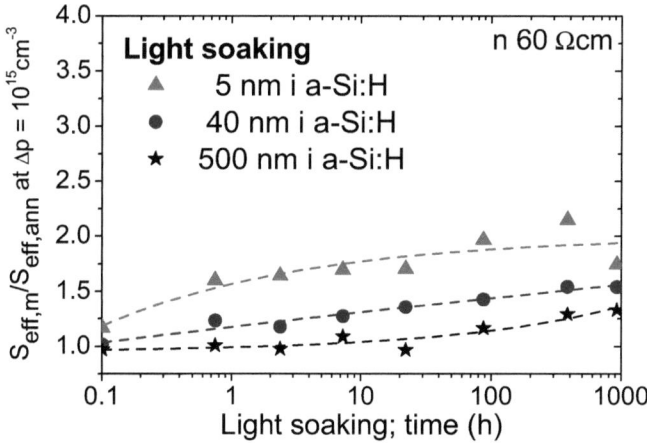

Figure 4.13: S_{eff} at $\Delta p = 1 \times 10^{15}$ cm^{-3}, normalized to its annealed value $S_{eff,ann}$ as a function of light soaking time for samples of the i a-Si:H passivation layers thickness series on lightly n-type doped c-Si, given in Fig. 4.11. Dashed lines are a visual guide.

Light soaking of the i a-Si:H passivated wafers shows small variations in the $S_{eff,m}$ values (Fig. 4.13). The largest variations are observed for the thinnest passivation layers, whereas variations are almost negligible for thick layers. This is in contradiction with the well known Staebler-Wronski effect observed in a-Si:H layers. In a-Si:H, the main metastable defect created by prolonged illumination is the silicon dangling bond, which acts as a recombination center. According to Stutzmann et al. [SJT85] the relative increase of N_{DB} exhibits a $t^{1/3}$-dependance on illumination time.

119

4.4. Intrinsic a-Si:H of varying thicknesses

By fitting our model to the experimental $S_{eff,m}(\Delta n)$-curve (Sec. 3.5.3), the surface DB density N_s at a given illumination time t_0 is evaluated and set to N_{s,t_0}. Prolonging the illumination duration, N_s should vary accordingly to

$$N_s(t) = N_{s,t_0} \times (t - t_0)^{2/9}, \qquad (4.2)$$

where the time is given in minutes. Figure 4.14 shows an example of a 500 nm thick a-Si:H i-layer on the lightly p-type doped c-Si wafer. With prolonged illumination time, increasing N_s values calculated from Eq. 4.2 should yield the higher $S_{eff,c}(\Delta n)$-curves shown by lines in Fig. 4.14. But none of the expected increase of S_{eff} is observed at all in the measured $S_{eff,m}(\Delta n)$-curves shown by stars in Fig. 4.14. However, this layer is thick enough to absorb significant light under light soaking conditions and should have suffered an increase in bulk dangling bond density, which does visibly not affect the value of $S_{eff,m}$.

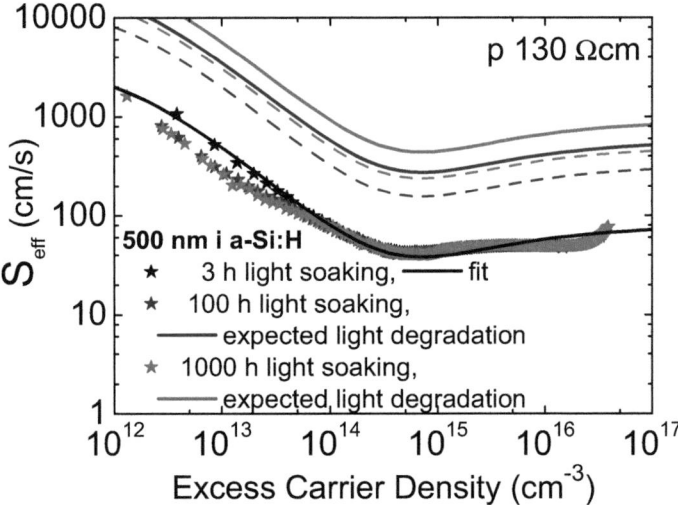

Figure 4.14: $S_{eff,m}(\Delta n)$-curves measured during light soaking of a thick (500 nm) i a-Si:H passivation layer. The stars show measured data while lines are calculated $S_{eff,c}(\Delta n)$-curves assuming $t^{1/3}$-increasing N_{DB} (bulk Staebler-Wronski effect). The dashed lines assume light degradation only on one side of the symmetrically passivated sample for the $S_{eff,c}(\Delta n)$-curve calculation, as only one side is illuminated in our experiment.

4.4. Intrinsic a-Si:H of varying thicknesses

4.4.3 Dark degradation

Because light soaking studies do not show the expected results, dark degradation is further investigated in order to elucidate the possible cause of the observed variations in $S_{eff,m}$ in the thinnest passivation layers. After 1000 h of light soaking, the initial value of $S_{eff,m}$ can be recovered by thermally annealing the samples again. Dark degradation is carried out in ambient atmosphere in a drawer at about 22 °C. Figure 4.15 shows the evolution of surface passivation properties as a function of time in the dark for the same layers as in Fig. 4.13. For direct comparison of dark degradation and light soaking, the results from Fig. 4.13 are represented again by the full symbols. Analogous to light soaking, the measured $S_{eff,m}$-evolutions are irrespective of the wafer doping type.

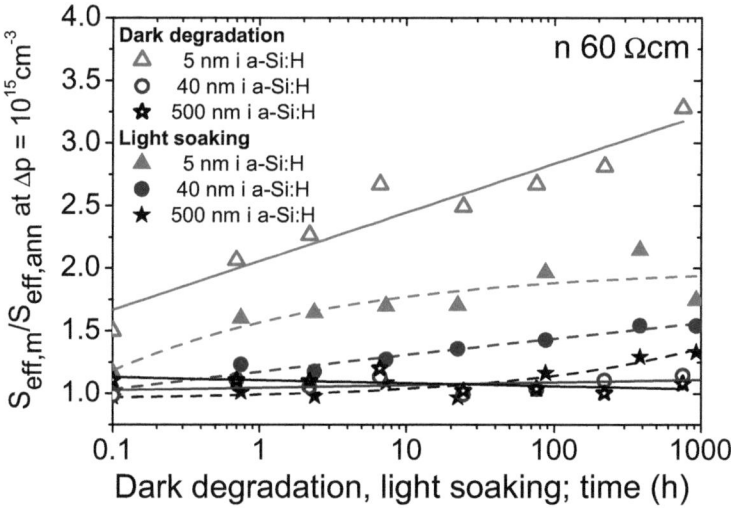

Figure 4.15: *Normalized $S_{eff,m}$ as a function of dark degradation time (open symbols and straight lines) for the i a-Si:H passivation layers of 3 different thicknesses shown in Fig. 4.13. For comparison, the light soaking behavior is represented again by solid symbols and dashed lines.*

From Fig. 4.15, it can be clearly seen that $S_{eff,m}$ of the thinnest a-Si:H i-layer degrades even more in the dark than under light soaking conditions. Thus, the light induced creation of metastable dangling bonds in the a-Si:H bulk is not the microscopic process governing the observed light and dark degradation kinetics as monitored by the variation of $S_{eff,m}$.

4.4. Intrinsic a-Si:H of varying thicknesses

Figure 4.12(a) shows the measured injection level dependent interface recombination of a sample consisting of 5 nm thick i a-Si:H layers grown symmetrically on a 60 Ωcm n-type doped c-Si wafer. Lifetime measurements are made directly after thermal annealing, as well as after 10 hours of dark storage in ambient air. The $S_{eff,m}(\Delta n)$-curve shape change in Fig. 4.12(a), corresponding to a change in the charge induced in the c-Si surface space-charge region, suggests that the degradation effect is related to a change at the outer surface of the i a-Si:H layer, like e.g. post-oxidation or water adsorption, leading to an uncontrolled change of the outer surface potential over time. This variation is reversible by thermal annealing. Indeed, we observe that the initial value of $\tau_{eff,m}$ can also be recovered by dipping the c-Si wafers, passivated with a-Si:H i-layers, in diluted HF.

These findings are not in accordance with those of Plagwitz et al. [PTB08] that observed an illumination induced passivation degradation while storage in the dark does not degrade their surface passivation. However, their 10 nm thin a-Si:H layers will be highly sensitive to an outer surface potential modification, e.g. by the formation of a native oxide. Storage in nitrogen atmosphere prevents the outer a-Si:H surface from oxidation and passivation is thus stable with respect to dark degradation time even for ultra-thin layers, in contrast to our dark degradation experiment carried out in ambient air. For an unambiguous verification of the light soaking/dark degradation/annealing/HF-dip experiment, it was repeated with optimized layers and fabrication procedures established, with the results shown in Fig. 4.16. In this case, 5 nm thick i a-Si:H layers are grown symmetrically on 3 Ωcm n-type doped c-Si, where an effective lifetime of 0.7 ms corresponding to an effective surface recombination velocity of 8.5 cm/s is measured after post-deposition annealing of two identical samples. Fig. 4.16 shows that interface recombination already more than doubles after only 25 min and increases by an order of magnitude after 15 h of storage in ambient atmosphere, the same for the sample stored under illuminated conditions as for the other one stored under dark conditions. The curve shape change in Fig. 4.17(a) suggests that the initially slightly negative surface charge on n-type c-Si prevailing from the amphoteric nature of silicon dangling bonds (Fig. 4.6 and related discussion in Sec. 4.3) is annihilated by the native oxide forming on top of the outer a-Si:H surface. Additionally, either this 5 nm thick i a-Si:H layer's (verified by HR-TEM micrographs to form a homogeneous dense layer) interface recombination is increased by the poorly passivated oxidized outer i a-Si:H surface or it increases because it is permeable to water absorption.

4.4. Intrinsic a-Si:H of varying thicknesses

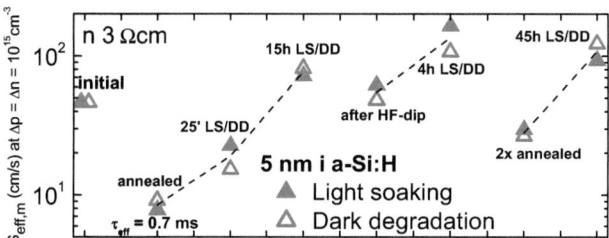

Figure 4.16: *Experimental re-verification under optimized deposition conditions of the effect of light soaking/dark degradation/thermal annealing/HF-dip on the passivation performance of 5 nm ultra-thin i a-Si:H layers symmetrically grown on 3 Ωcm n-type c-Si. Two identical samples are prepared for the simultaneous study of light and dark degradation. $S_{eff,m}$-values measured at $\Delta p = 1 \times 10^{15}$ cm^{-3} are displayed after thermal annealing, consecutive parallel light/dark degradation, then HF-dip and renewal parallel light/dark degradation and finally a second thermal anneal and a third parallel light/dark degradation.*

In contrast to the previous experiment, with a lower base passivation performance, the HF-dip performed to remove the native oxide formed on top of the outer i a-Si:H surface only slightly recovers the degraded interface passivation performance as shown at an injection level of $\Delta p = 1 \times 10^{15}$ cm^{-3} in Fig. 4.16. Figure 4.17(b) then confirms, by the recovery of the initial $\tau_{eff,m}(\Delta n)$ curve shape, that the native oxide on top of the a-Si:H could be stripped off by the HF-dip, although the interface dangling bond density could not be decreased. 4 h after the HF-dip, the a-Si:H surface is already considerably re-oxidized (Figs. 4.16 and 4.17(b)). Whereas for the previous experiment, successive thermal annealings permitted an almost perfect recovery of the freshly annealed passivation performances, in this case a thermal annealing permits only a partial re-establishment of the low interface state density of the freshly annealed state (Fig. 4.16), but the characteristic curve shape of the non-oxidized a-Si:H surface is recovered (Fig. 4.17(c)). It has to be noted that $S_{eff,m}$ is still lower than the lowest measured in the previous experiment (30 cm/s vs 80 cm/s). Finally, the third set of parallel light and dark degradation confirms the results of the first and the second one (again Figs. 4.16 and 4.17(c)). Although the reproducibility of the conditions for such experiments is low, it can be concluded that the resistance of ultra-thin a-Si:H passivation performances against degradation in ambient atmosphere is not related to illumination but to the effect of the ambient atmosphere.

4.4. Intrinsic a-Si:H of varying thicknesses

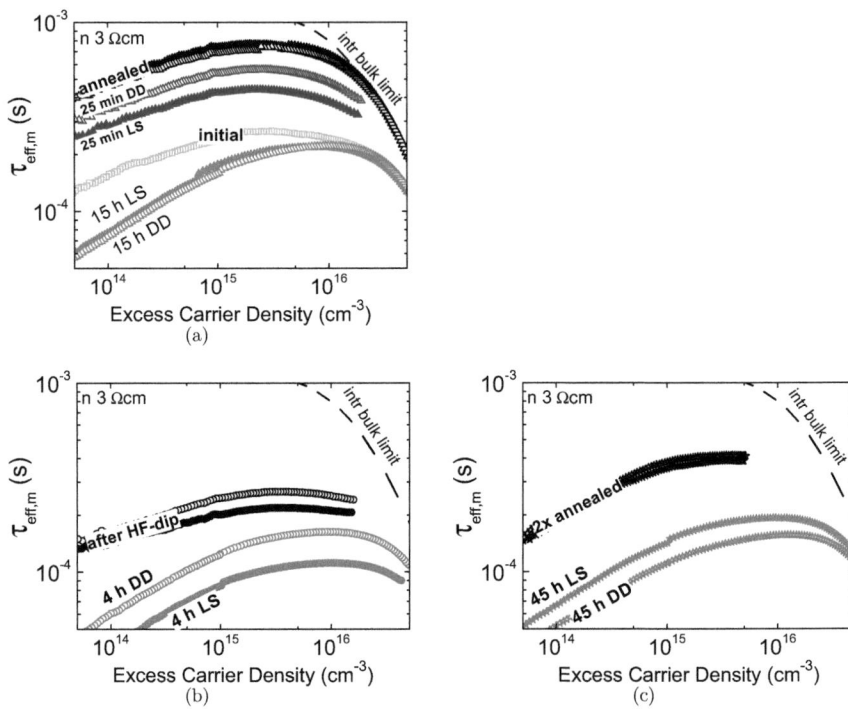

Figure 4.17: *Injection level-dependance of the lifetime curves measured on n 3 Ωcm n-type c-Si passivated by 5 nm i a-Si:H a) initially, after post-deposition thermal annealing and consecutive light/dark degradation, b) consecutive HF-dip and again light/dark degradation and finally c) renewal thermal annealing followed by light/dark degradation. Note that for better readability the lifetime scales are reduced with respect to the other $\tau_{eff}(\Delta n)$-plots in this study.*

Because of their instability with respect to storage in air, such ultra-thin i a-Si:H layers, unprotected by further grown capping layers (such as a doped a-Si:H layer in heterojunction solar cells and a SiN_x or a SiO_x layer in a c-Si solar cell [PTTB06, HSK$^+$08]) are of no practical use. Thus, typical uncapped i a-Si:H passivation layers used in this work to study the a-Si:H/c-Si interface recombination mechanism are about 45 nm thick. Their good interface passivation performance on various flat c-Si wafers was discussed in Sec. 4.3. Such thicker layers permit the suppression of the post-oxidation effects on the a-Si:H/c-Si interface passivation.

4.5 Additional field-effect passivation

Microdoping i a-Si:H passivation layers or fixing the i a-Si:H passivation layers' outer surface potential by a doped a-Si:H layer permits the addition of field-effect passivation to the interface defect density reduction of a solely i a-Si:H passivation. This section first qualitatively discusses the injection level-dependance of experimental $S_{eff,m}(\Delta n)$-curves and offers thus a deeper understanding of the a-Si:H passivation mechanism on c-Si. Modeling emitter and BSF layer stack passivations then permits a quantitative evaluation of these layers' performances for further integration in complete Si HJ solar cells.

The same i a-Si:H passivation layers used for degradation studies in Sec. 4.4, with not yet established optimized layers and fabrication procedures, have been used for the evaluation of field-effect passivation. Therefore, the relatively high values of $S_{eff,m}(\Delta n)$ presented here were obtained at the beginning of this work (comment in Sec. 4.4.1). While the standard i a-Si:H layer deposition conditions for this PECVD-chamber (Sec. 2.3.2) date from before this work, deposition conditions for highly conductive doped a-Si:H/µc-Si:H layers had to be developed. The deposition conditions used here for the first test-structures incorporating doped a-Si:H/µc-Si:H layers are not optimal in comparison to those developed later on. Results with optimized doped layers (fabrication described in Sec. 2.3.4) are presented at the end of Sec. 4.5.2.

4.5.1 Microdoped a-Si:H

Microdoping (µdop) varies the average charge state of DBs in i a-Si:H without the simultaneous increase in the a-Si:H's dangling bond density that is introduced by heavier doping such as needed for the fabrication of doped a-Si:H layers [Str85]. Thus microdoping permits the introduction of additional field-effect passivation without increasing the interface dangling bond density (exact growth conditions are given in Sec. 2.3.3). Figure 4.18 shows the injection level-dependance of $S_{eff,m}$ for the lightly n-type doped (60 Ωcm) c-Si wafer passivated with n- and p-type microdoped a-Si:H layers having a thickness of 40 nm. For direct comparison, the $S_{eff,m}(\Delta n)$-curve of the 40 nm thick fully intrinsic a-Si:H passivation layer is displayed again.

4.5. Additional field-effect passivation

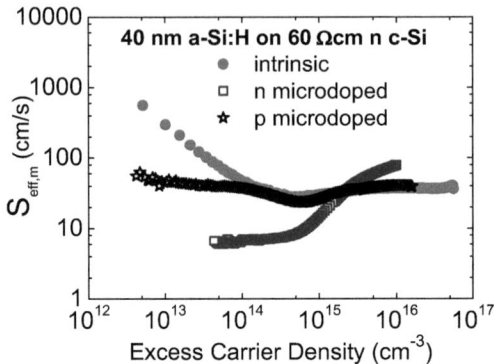

Figure 4.18: *Passivating effect of n- and p-type microdoped a-Si:H layers compared to i a-Si:H on lightly n-type doped c-Si (60 Ωcm).*

A sketch of the recombination model pertaining to this case at the a-Si:H/c-Si interface is given in Fig. 4.19 for intrinsic c-Si. Compared to i a-Si:H passivation, p-type microdoping leads to a decreased interface recombination for low injection levels, while the rest of the $S_{eff,m}(\Delta n)$-curves overlap (see Fig. 4.18). n-type microdoping completely modifies the injection level-dependance of $S_{eff,m}$ as compared to i a-Si:H passivation, such that the corresponding $S_{eff,m}(\Delta n)$-curves cross (see Fig. 4.18, injection level of $\Delta p = \Delta n = 2 \times 10^{15}$ cm^{-3}).

Figure 4.19: *Sketch of the average state of charge of the recombination centers in microdoped a-Si:H layers passivating intrinsic c-Si: a) n-type µdop and b) p-type µdop (charge density is adapted to measurements). The sketched occupation level in the a-Si:H distant from the interface in the drawings illustrate the expected bulk occupation in microdoped a-Si:H layers.*

4.5. Additional field-effect passivation

The $S_{eff,m}(\Delta n)$-curve shape changes introduced by microdoping, starting from the i a-Si:H case in Fig. 4.18, are typical for the addition of a small positive field-effect passivation and a negative field-effect passivation on lightly n-type doped c-Si. This can be explained as follows: once the recombination centers capture cross-section ratios are set, the shape of an $S_{eff,m}(\Delta n)$-curve is solely given by the values of the surface carrier densities n_s and p_s. The latter are fixed by the surface potential ψ_s that is itself a function of the surface charge density Q_s, the wafer doping n_0 (p_0) and the injection level Δn, calculated by numerically solving the non-linear Eq. 3.24. The definitions of electron energy and potentials are given in Fig. 3.5. Figure 4.20(a) shows $\psi_s(\Delta n)$ for our lightly n-type doped c-Si wafer (60 Ωcm $\sim n_0 = 1 \times 10^{14}$ cm^{-3}) with a slightly positive ($Q_s = +1 \times 10^{10}$ cm^{-2}) and a stronger negative ($Q_s = -10 \times 10^{10}$ cm^{-2}) surface charge density. The sign of ψ_s is given by the sign of Q_s, and its magnitude decreases with increasing injection level (Fig. 4.20(a)). For n-type c-Si, the calculation of n_s and p_s from ψ_s in Eq. 3.20 reduces to $n_s = (n_0 + \Delta n)e^{q\psi_s/kT}$ and $p_s = \Delta n e^{-q\psi_s/kT}$, where for high injection level conditions, $n_s(\Delta n \gg n_0 = 1 \times 10^{14}$ cm$^{-3}) = \Delta n e^{q\psi_s/kT}$ and for low injection level conditions $n_s(\Delta n \ll n_0 = 1 \times 10^{14}$ cm$^{-3}) = n_0 e^{q\psi_s/kT}$. The small positive surface charge density induced by p-type microdoping of the i a-Si:H passivation layer (Fig. 4.19(b)) leaves $\psi_S \approx 0$V at high injection levels and therefore $n_s = p_s$. This is the same situation without any surface charge and hence the curves of both surface and both bulk carrier densities in Fig. 4.20(b) overlap ($n_s = p_s = n_b = p_b$). At low injection levels, the small positive surface charge increases the ratio of n_s/p_s as compared to the flatband case (Fig. 4.20(b)) and therefore recombination is reduced as compared to the flatband case due to the reduced density of hole recombination partners for the more numerous electrons at the interface. This is verified by the decreased interface recombination measured at low injection levels on the p-type microdoped passivation layer in Fig. 4.18. The n-type microdoping leads to an increased average negative state of charge of the dangling bonds within the passivation interface as schematically shown in the sketch of Fig. 4.19(a). For a lightly n-type doped c-Si wafer, already a medium negative surface charge produces inversion conditions in the c-Si surface as not much negative charge is needed to repel the already sparse electrons. Again, the curve shape change observed in Fig. 4.18 can be explained by means of the surface carrier densities displayed in Fig. 4.20(c). From Fig. 4.20(a) we see that ψ_s is only negligible for very high injection level conditions. For $\Delta n > n_0$, ψ_s is already negative and therefore $p_s(\Delta n \gg 1 \times 10^{14}$ cm$^{-3}) = \Delta n e^{-q\psi_s/kT} > n_s(\Delta n \gg 1 \times 10^{14}$ cm$^{-3}) =$

4.5. Additional field-effect passivation

$\Delta n e^{q\psi_s/kT}$ and because $\sigma_p^0 > \sigma_n^0$, recombination is maximal, as illustrated in the surface plot of the recombination rate in Fig. 3.18. For decreasing Δn, ψ_s becomes increasingly negative, i.e. $p_s(\Delta n < 1 \times 10^{14} \text{ cm}^{-3}) = \Delta n e^{-q\psi_s/kT} \gg n_s(\Delta n < 1 \times 10^{14} \text{ cm}^{-3}) = 1 \times 10^{14} e^{q\psi_s/kT}$ (Fig. 4.20(c)). The presence of very few interface electrons, as compared to the more numerous holes, strongly limits the interface recombination at lower injection levels. Both, the higher recombination rate at high injection levels, as well as the low recombination rate at low injection levels are verified by the $S_{eff,m}(\Delta n)$-measurement of the n-type microdoped passivation layer as compared to its fully intrinsic counterpart in Fig. 4.18.

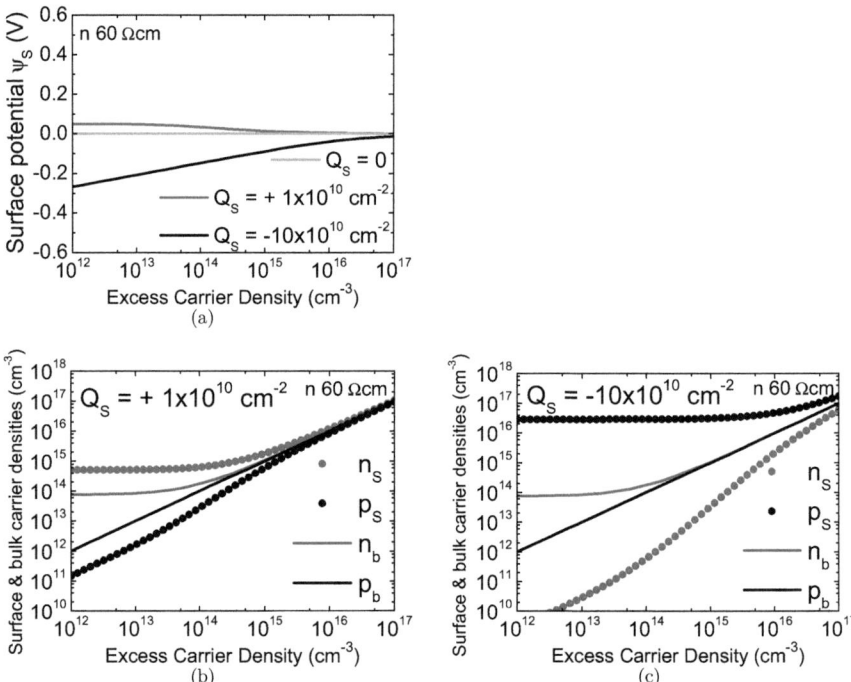

Figure 4.20: *a) Calculation of the injection level dependent surface potential $\psi_s(\Delta n)$ from a slightly positive and a stronger negative surface charge density Q_s on lightly n-type doped c-Si (Eq. 3.24). b) and c) Calculated surface carrier densities (from Eq. 3.20) including the flatband case ($n_s = n_b = n_0 + \Delta n$, $p_s = p_b = p_0 + \Delta p$) for b) a slightly positive and c) a stronger negative surface charge density (Eq. 3.20).*

4.5. Additional field-effect passivation

In conclusion, we observe that, on one hand, p-type microdoping induces a small additional field-effect passivation while, on the other hand, n-type microdoping is sufficiently high to invert the c-Si surface. The chosen level of microdoping is low enough not to induce additional interface dangling bonds.

Remark In fact, when growing passivating n-type microdoped layers thicker and thinner than 40 nm, the optimum layer thickness is shifted to higher values with respect to the intrinsic case (Fig. 4.11), as shown in the graphical comparison of the corresponding $S_{eff,m}(\Delta n)$-curves in Fig. 4.21.

For the 5 nm thick n-type microdoped layer, Fig. 4.21 shows curve segments consisting of generalized QSS (medium to high illumination levels) and curve segments consisting of transient (low to medium illumination levels) Sinton lifetime measurements, including a fit to the generalized QSS curve with $N_s = 5.5 \times 10^{10}$ cm^{-2} / $Q_s = -11 \times 10^{10}$ cm^{-2}. Discrepancies in the transient and the generalized QSS mode become obvious and fitting to such curves is ambiguous and thus is not attempted.

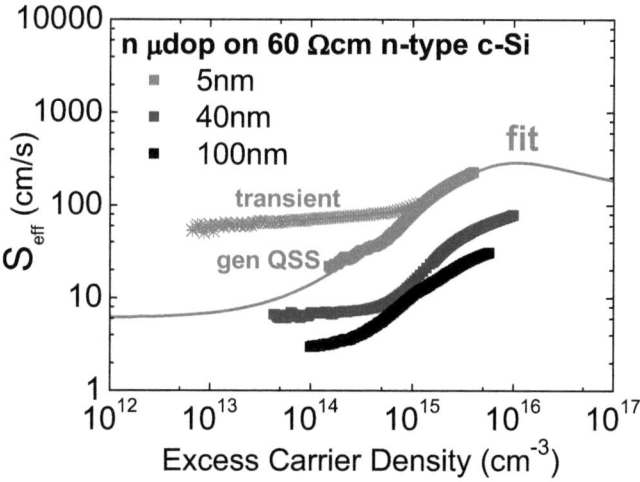

Figure 4.21: *Thickness dependent surface passivation of n-type microdoped i a-Si:H layers on lightly n-type doped c-Si. For the thinnest passivation layer, $\tau_{eff,m}(\Delta n)$-curve measurements made in the transient and the generalized QSS modes differ.*

4.5. Additional field-effect passivation

4.5.2 Stacks of intrinsic a-Si:H plus doped a-Si:H/μc-Si:H

While corona surface charging of SiO_2 passivation layers permits the tuning of field-effect passivation for examination purposes [GBRW99], the deposition of a doped a-Si:H overlayer can produce such a field-effect passivation permanently. This is illustrated in Fig. 4.22, which shows the injection level-dependance of the surface recombination velocities of symmetrically grown stacks consisting of 10 nm i a-Si:H plus 30 nm doped a-Si:H that passivate 60 Ωcm n-type c-Si. For comparison purposes the intrinsic passivation's $S_{eff,m}(\Delta n)$-curve is shown again too.

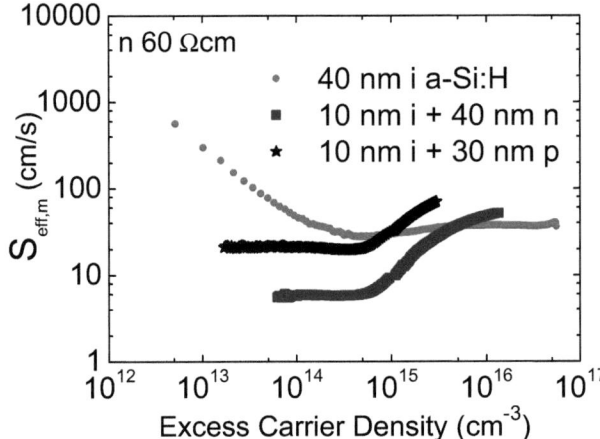

Figure 4.22: *Passivating effect of i/n-doped and i/p-doped stacks of a-Si:H layers compared to i a-Si:H passivation on lightly n-type doped c-Si (60 Ωcm).*

In a similar manner as in the case of microdoping i a-Si:H layers (Sec. 4.5.1), these intrinsic/doped stacked passivation schemes permit to shift the Fermi level in the passivating i a-Si:H layers, i.e. to modify the average state of charge of the recombination centers, without greatly increasing the interface dangling bond density. This is shown for comparison with the i a-Si:H single layer passivation in Fig. 4.22. 10 nm thick i a-Si:H layers with interface relaxing capping layers reach low N_s. The doped layers are grown up to a thickness of several tens of nanometers and physically fix the i a-Si:H potential.

4.5. Additional field-effect passivation

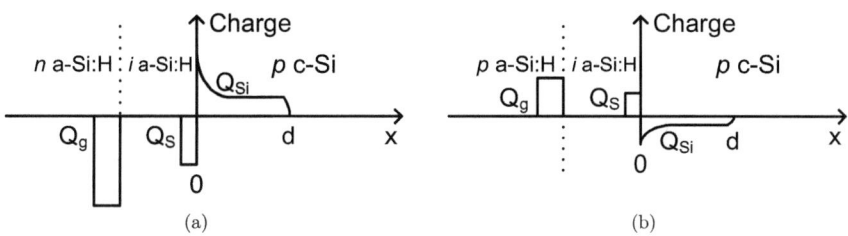

Figure 4.23: *Sketch of the charge distribution when passivating lightly p-type doped c-Si by stacks of a) i/n a-Si:H and b) i/p a-Si:H layers.*

Figure 4.23 shows this by means of a sketch of the prevailing charge density distribution. The resulting Fermi level shift in the i a-Si:H layer changes the average charge state of dangling bonds in the same manner as in the case of microdoping.

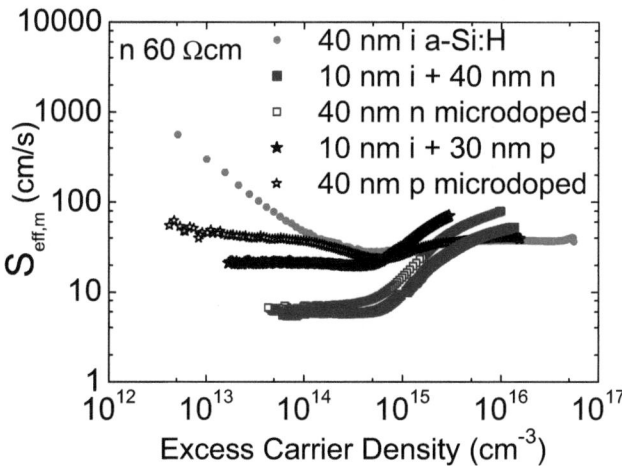

Figure 4.24: *Passivating effect of i/n-doped and i/p-doped a-Si:H layer stacks (full symbols) compared to n- and p-type microdoped a-Si:H layers (open symbols), as well as to i a-Si:H (dots), on lightly n-type doped c-Si (60 Ωcm). The upward shift of the Fermi level in a-Si:H obtained either from n-type microdoping or capping by a n-doped layer results in almost the same injection level-dependance of $S_{eff,m}$ (square symbols). This effect is in tendency also observed for a downward shift of the Fermi level (star symbols).*

4.5. Additional field-effect passivation

Fig. 4.24 clearly illustrates the equivalent effect of stack deposition and microdoping by reproducing, for comparison purposes, the data from passivation with microdoped layers (open symbols). The $S_{eff,m}(\Delta n)$-curve shape corresponding to the i/p a-Si:H-layer stack differs from the p-type microdoped passivation layer one. This is because this doped p-type layer introduces a larger Fermi level shift in the underlying thin i a-Si:H layer than our p-type microdoping does. Therefore, the c-Si surface is in accumulation and recombination at low injection levels is reduced by the presence of very few interface holes as compared to the very numerous electrons (equivalent discussion in Sec. 4.5.1). Contrariwise, at higher injection levels recombination is increased due to an increased interface dangling bond density related to the presence of the p a-Si:H layer, that is also observed by other authors [DWB06, GRB+05]. As mentioned previously, the deposition parameters of these doped layers do not yet correspond to the optimized growth conditions described in Sec. 2.3.4. Further on we will see that optimized p-type layers, growing at the transition of a-Si:H to μc-Si:H, have a less detrimental effect on the surface dangling bond density. Compared to n-type microdoping, an i-layer capping n-type overlayer yields a slightly increased field-effect passivation.

In the aim of characterizing lifetime test structures to forecast the performance of configurations eventually used in Si heterojunction solar cells, we grow symmetrical stacks of i a-Si:H / doped a-Si:H/μc-Si:H layers in device layer thicknesses, i.e. 5 nm i a-Si:H + 15 nm doped a-Si:H/μc-Si:H, and on more highly doped n-type c-Si. All results shown further on are based on the optimized p- and n-type a-Si:H/μc-Si:H layers described in Sec. 2.3.4 and are processed under the optimized fabrication procedures shown to reach high measured lifetimes (Sec. 4.3). Figure 4.25 compares the passivation performance of symmetrically grown emitter (i/p) and BSF (i/n) layer stacks on 2.8 Ωcm n-type doped c-Si with the one of 45 nm thick i a-Si:H. Injection level dependent recombination is again shown by means of $\tau_{eff,m}(\Delta n)$-curves, as the bulk c-Si lifetime limit is reached and even exceeded, and thus no useful $S_{eff,m}$ values can be extracted at high injection levels.

4.5. Additional field-effect passivation

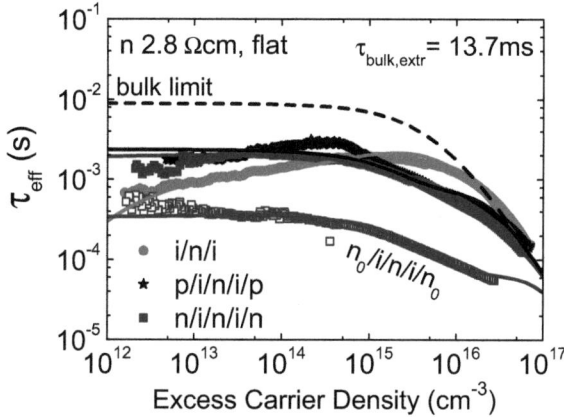

Figure 4.25: *Passivation performance of symmetrically grown emitter (i/p) and BSF (i/n) layer stacks compared to i a-Si:H passivation on n-type doped c-Si of a higher doping level such as that used also for Si heterojunction solar cell fabrication (2.8 Ωcm). Symbols show lifetime measurements while lines are fits made with the model parameters N_s and Q_s from Tab. 4.3.*

The corresponding fit values in Tab. 4.3 giving the best accordance with the experimental (symbol) curves show the expected behavior for stacks of i/p and i/n a-Si:H layers; that is, on one hand, Q_s is strongly increased and carries the sign of the doped layer's doping and on the other hand N_s is slightly increased due to the small distance of the more defective doped layer to the c-Si interface.

n 2.8 Ωcm c-Si	N_s (10^9 cm^{-2})	Q_s (10^{10} cm^{-2})
45nm i a-Si:H	1.0	-2.2
5nm i a-Si:H + 17 nm p a-Si:H/μc-Si:H	2.3	+55
5nm i a-Si:H + 17 nm n_0 a-Si:H/μc-Si:H	35	-18
5nm i a-Si:H + 17 nm n a-Si:H/μc-Si:H	5	-18

Table 4.3: *Model parameters N_s and Q_s giving the best agreement between the measured and calculated $\tau_{eff}(\Delta n)$-curves in Fig. 4.25.*

4.5. Additional field-effect passivation

The experimental open square curve shows the passivation performance of our i/n_0 layer stack, that is our i/n layer stack such as initially developed on glass. Although i/n_0 ensures field-effect passivation, it also induces detrimental interface defects. By fine-tuning this i/n layer stack's passivation on c-Si (final process parameters in Sec. 2.3.4), field-effect passivation can be maintained while reducing the interface dangling bond density by almost an order of magnitude (compare in Tab. 4.3 N_s of i/n_0 and i/n). Such symmetrical a-Si:H layer passivations are thus a fast diagnostic procedure to evaluate the suitability of pre-cleanings, preconditionings and of layer stacks for integration in a-Si:H/c-Si heterojunction solar cells, to be discussed in Sec. 5.3.

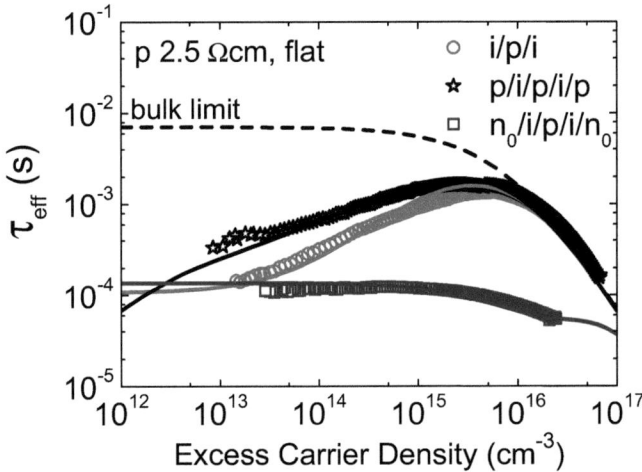

Figure 4.26: *Passivation performance of symmetrically grown emitter (i/n_0) and BSF (i/p) layer stacks compared to i a-Si:H passivation on 2.5 Ωcm p-type doped c-Si. Symbols show measurement data and lines are fits.*

On p-type c-Si, the same i-layer, i/p- and i/n_0 layer stack passivation [FOVS+07] lead to lower low injection level lifetimes (Fig. 4.26). Fitting to these experimental curves and plotting them (together with the ones on n-type c-Si from Fig. 4.25) on the surface recombination rate's surface plot in Fig. 4.27 illustrates that the lower lifetimes at low injection levels on p-type (open symbols) than on n-type (full symbols) c-Si result from the slight neutral capture cross-section asymmetry in favor of the hole capture cross-section, i.e. $\sigma_p > \sigma_n$ leading in general to higher recombination when $p_s >$

4.5. Additional field-effect passivation

n_s. In Sec. 4.8 we discuss the consequences of this less favorable injection level-dependance on p-type rather than on n-type c-Si for heterojunction solar cell fabrication.

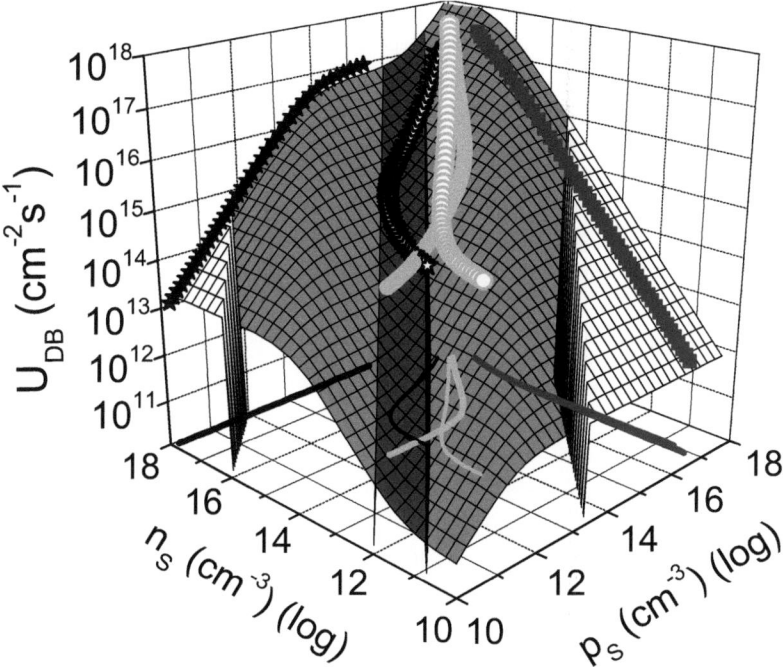

Figure 4.27: *Trajectories over the surface plot of $U_{DB} = \mathrm{f}(n_s, p_s)$ corresponding to the fitted temporal variation of n_s/p_s in 2.8 Ωcm n- (full symbols) as well as 2.5 Ωcm p-type c-Si (open symbols) c-Si passivated with the i-layer, i/p- and i/n layer stacks in Figs. 4.25 and 4.26. For better visualization, the interface dangling bond density N_s is set to the same value for better visualization.*

Finally, asymmetrically doped stacks of intrinsic/doped layers are grown on both sides of more highly doped n- and p-type c-Si wafers. These structures result in heterojunction solar cells, simply by further transparent contact deposition. As we cannot distinguish between front and back surface recombination, such samples are only co-deposited with a-Si:H/c-Si heterojunction solar cells to serve as control samples for monitoring possible solar cell V_{OC}-losses due to the front transparent contact (TCO) (Sec. 5.3).

4.6 Atomic structure of the a-Si:H/c-Si heterointerface

The importance of a crystallographically abrupt a-Si:H/c-Si heterointerface is pointed out in this section by the comparison of HR-TEM micrographs and lifetime measurements. The preferential epitaxial growth of a-Si:H on <100> over <111> oriented c-Si already observed by other researchers is confirmed.

Previous amorphous/crystalline Si interface recombination studies also focus on its atomic nature [YPW+06, WIP+05, GvdOH+08]. Best passivation necessitates an abrupt and flat interface of the a-Si:H layer to c-Si [WIP+05, DWK07, DBL+08]. Our standard i a-Si:H layer perfectly fulfills this criteria as shown by the HR-TEM micrograph (Sec. 2.1.2.3) in Fig. 4.28. The growth conditions of the i-layers presented in this section are given in Sec. 2.3.2.

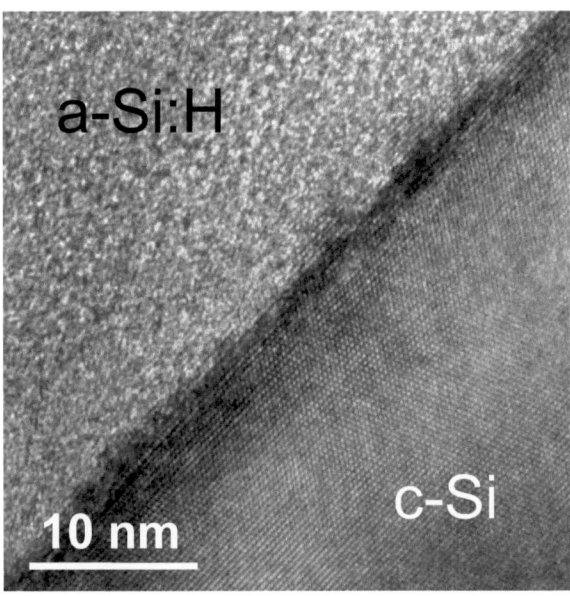

Figure 4.28: *HR-TEM micrograph of the a-Si:H/c-Si heterointerface showing its crystallographic abrupt interface. The standard (hydrogen dilution 2.7) i a-Si:H/<100> c-Si interface shows the same microscopic picture than the highly hydrogen diluted ($H_{dil} = 9$) i a-Si:H/<111> c-Si interface.*

4.6. Atomic structure of the a-Si:H/c-Si heterointerface

The very first passivation layers were grown without any hydrogen dilution to prevent epitaxial growth on c-Si. Having checked that also our low bulk-defect (dangling bond) density a-Si:H layer grows abruptly amorphous on c-Si despite its hydrogen dilution, this device-quality layer was later used as the standard layer (hydrogen dilution ~ 2.7 corresponding to silane concentration $\sim 27\%$). Note that under such a deposition temperature and dilution other researchers usually report epitaxial growth. As a consequence, a-Si:H layers are commonly deposited on c-Si at a lower temperature and without hydrogen dilution. Possibly the high deposition rate of VHF-PECVD as compared to lower frequency PECVD [CWS87] is the reason for the larger process window we met for abrupt amorphous growth.

Epitaxial growth of the a-Si:H passivation layer is much more easily observed on <100> than on <111> oriented c-Si [WIP+05, DBL+08]. An i a-Si:H layer deposited under a high hydrogen dilution of 9 (corresponding to a silane concentration of 10%) and therefore, a reduced growth rate from 3 Å/s to under 2 Å/s, indeed yields abrupt amorphous growth on <111> oriented c-Si while providing initial epitaxial growth on <100> oriented c-Si, compare Fig. 4.28 and Fig. 4.29.

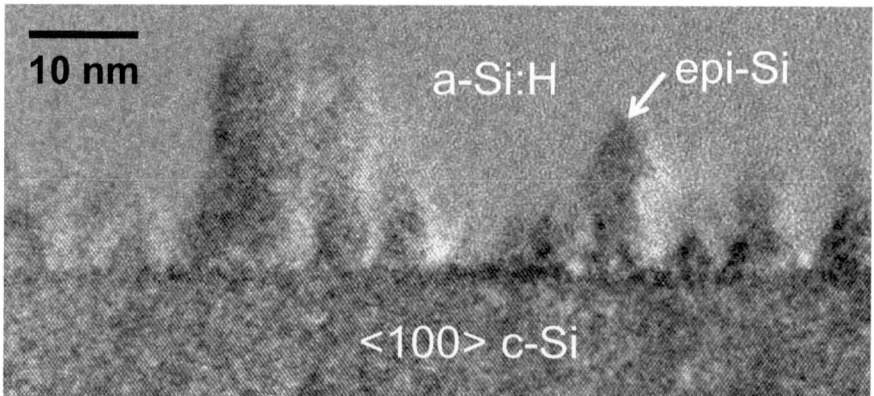

Figure 4.29: *Initial epitaxial growth of a highly hydrogen diluted i a-Si:H layer ($H_2/SiH_4 = 9$) on <100> oriented c-Si. Under the same growth conditions, this layer grows fully amorphous on <111> oriented c-Si, yielding an abrupt crystallographic interface (Fig. 4.28).*

Such ill-defined crystallographic interfaces between amorphous and crystalline Si result in a interface recombination increase as shown by the life-

4.6. Atomic structure of the a-Si:H/c-Si heterointerface

time drop in Fig. 4.30(a) compared to the fully amorphous layer on the same <100> oriented wafer. Note, that this decrease in lifetime is not due to a possibly poorer interface defect reduction capability of the higher diluted a-Si:H layer, as its passivation performance on <111> oriented c-Si is the same as the one of the less diluted standard i a-Si:H layer, but may result from the increased developed area of the crystallographic interface between a-Si:H and the epitaxial crystalline interface. Additionally, one can recognize in the HR-TEM micrograph in Fig. 4.29 a large amount of crystallographic defects (twins) in the epitaxially grown part of the i a-Si:H layer. Such crystallographic stacking faults probably contain a large amount of defects similar to the unpassivated c-Si surface. Thus, as deduced from the $S_{eff,m}(\Delta n)$-curves in Fig. 4.30(b), the initially epitaxial growth results in an interface defect density increase by a factor of 40. As already discussed (Sec. 3.4.1, i.e. Fig. 3.9), high injection level $S_{eff,m}(\Delta n)$-curves are affected by uncertainties. For comparison purposes, we therefore adopt the $S_{eff,m}$ value at $\Delta n = 5 \times 10^{15}$ cm^{-3} (before the strong decrease of $S_{eff,m}$) as the minimal $S_{eff,m}$ values of these passivations.

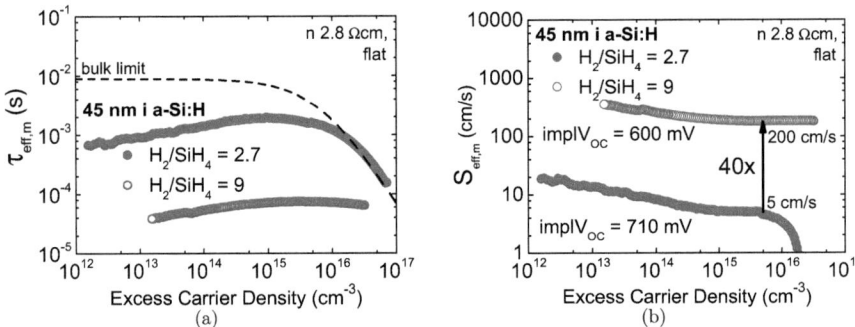

Figure 4.30: *Influence of the a-Si:H/c-Si heterointerface's crystallographic nature on interface recombination: comparison of the passivation performance of a fully amorphous i-layer ($H_{dil} = 2.7$, HR-TEM micrograph in Fig. 4.28) and an initially epitaxially growing i-layer ($H_{dil} = 9$, HR-TEM micrograph in Fig. 4.29). a) Lifetime decrease corresponding to b) an effective surface recombination velocity increase by a factor of 40.*

Note here that an abrupt microcrystalline growth on c-Si as shown in Fig. 4.31 yields the same excellent measured carrier lifetimes as does an abruptly amorphous grown i-layer.

4.6. Atomic structure of the a-Si:H/c-Si heterointerface

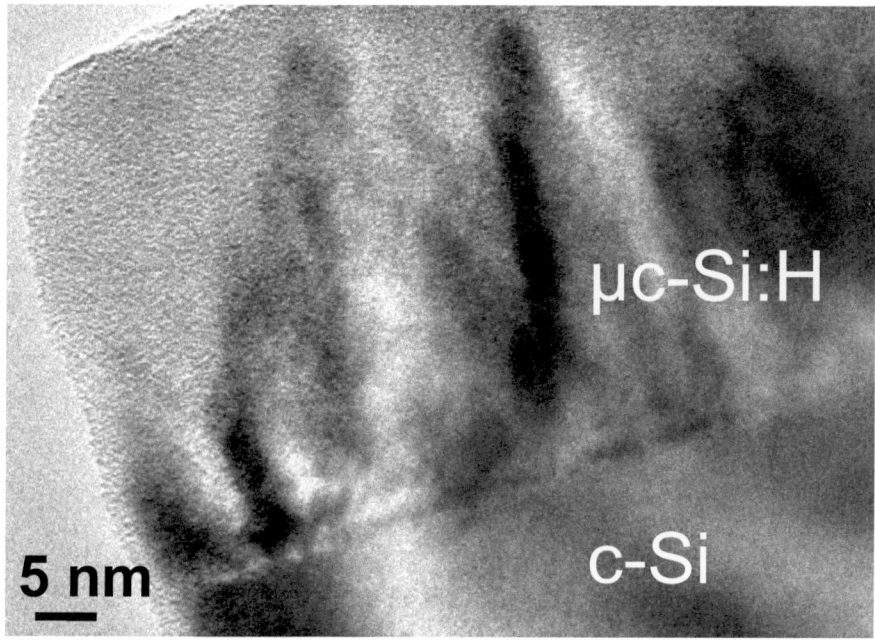

Figure 4.31: *Abrupt microcrystalline growth on c-Si yielding the same excellent measured lifetimes as abruptly amorphous grown i-layers (Fig. 4.28).*

On textured c-Si, as shown in Fig. 4.32, the highly diluted i a-Si:H layer yields an abrupt amorphous growth on the <111> oriented pyramidal facets, as expected from the abrupt amorphous growth on flat <111> oriented c-Si. In the pyramidal valley, a small epitaxial structure can be found. Not surprisingly, also on textured c-Si, a lifetime drop is observed in Fig. 4.33(a) as compared to the less diluted standard i a-Si:H passivation layer. Because a badly defined interface is only present in pyramidal valleys, the corresponding interface state density increases only by a factor of 4 here ($S_{eff,m}(\Delta n = 10^{15}$ cm$^{-3})$ in Fig. 4.33(b)) as compared to the one of the completely epitaxial interface on <100> oriented c-Si shown in Fig. 4.30(b) (increase of the interface state density by a factor of 40).

4.6. Atomic structure of the a-Si:H/c-Si heterointerface

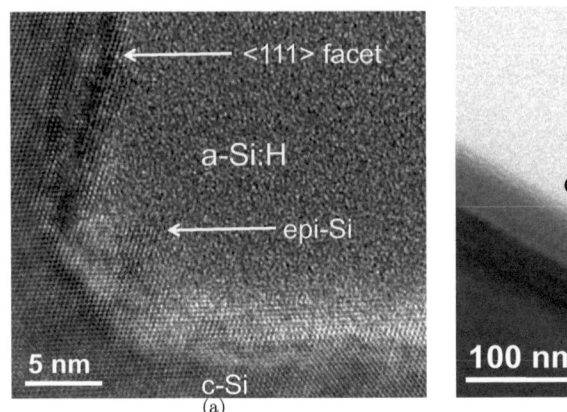

Figure 4.32: *Growth of a highly diluted i a-Si:H layer on pyramidally textured c-Si. a) On the <111> oriented facets abrupt amorphous growth prevails as expected from the abrupt amorphous growth on flat <111> oriented c-Si. In the pyramidal valley, a small epitaxial structure can be found. a) and b) show two different pyramidal valleys. b) a-Si:H layer growth is conformal with respect to the sharp c-Si valley. The absence of a protuberance in Fig. b) indicates that the small crystalline structure within the amorphous layer in Fig. a) results from epitaxial growth and that this feature does not belong to the original substrate (compare to Fig. 4.34 showing the case of an a-Si:H layer growing conformal on a mini-pyramid occupying a large pyramidal valley). The small crack visible in the pyramidal valley of Fig. b) additionally shows that the growth front collision of the amorphous layer yields inhomogeneities in the layers' density in the valley's axis. Figure a) and Fig. b) show two different pyramidal valleys.*

Looking at the SEM micrograph in Fig. 4.38, where an initial wafer surface populated with very small pyramids is observed, could lead one to the conclusion that the small crystalline structure observed in Fig. 4.32 corresponds to a nanometrical mini-pyramid and is thus already present before the *i* a-Si:H layer's growth. But the HR-TEM micrograph showed no such nanopyramid at the oxidized, textured c-Si surface where, after oxide etch and subsequent VHF-PECVD thin-film Si layer deposition, small crystalline structures were found (Fig. 5.18). In addition, the SEM micrograph in Fig. 4.34 shows the conformal coverage of a mini-pyramid by *i* a-Si:H leading to an a-Si:H bump in the pyramidal valley while the small crystalline structure in Fig. 4.32(b) does not modify the replication of the

4.7. Influence of the texture morphology

basically sharp pyramidal valley by the i a-Si:H layer's outer surface.

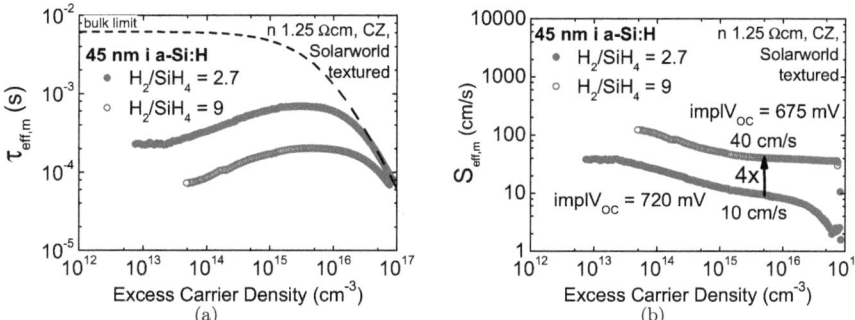

Figure 4.33: *Influence of the textured a-Si:H/c-Si interface's nature on the interface recombination rate: comparison of the passivation performance of an i a-Si:H passivation layer growing under high hydrogen dilution conditions ($H_{dil} = 9$) showing a small epitaxial structure in the pyramidal valley on the TEM micrograph (Fig. 4.32), and the standard i a-Si:H layer grown within a lower hydrogen dilution ($H_{dil} = 2.7$) on the same textured wafer. a) Lifetime drop due to epitaxy in pyramidal valleys corresponding to b) an effective surface recombination velocity increase by a factor of 4.*

Finally, if the small crystalline structure in the pyramid valley would be a nanopyramid contained in the c-Si substrate, less surface passivation deterioration as observed in Fig. 4.33 would be expected as the highly diluted i a-Si:H layer excellently passivates <111>-oriented flat c-Si. The small crystalline structure in the pyramidal valley shown in Fig. 4.32 thus prevails from initially epitaxial growth of the highly diluted a-Si:H layer induced therein by stress and/or the different c-Si orientation.

4.7 Influence of the texture morphology

The use of textured c-Si is indispensable to achieve of the most efficient c-Si solar cells. In this section we study thus interface recombination at the textured c-Si/a-Si:H interface, starting from thicker i a-Si:H passivation up to symmetrical thin-film Si emitter and BSF passivation in view of Si HJ solar cell formation. Although the requirements on the textured surface's morphology are high, excellent results can be reached by minimiz-

4.7. Influence of the texture morphology

ing the previously identified growth problems in the textured c-Si's valleys.

To enhance photogeneration within the c-Si bulk by minimizing optical reflection losses, c-Si solar cell surfaces are not flat but textured (Sec. 2.4.1, especially Figs. 2.16 and 2.17). In the past, the passivation of textured surfaces has shown to be more challenging than the one of flat surfaces: textured c-Si wafers have typically shown an increase in the surface recombination velocity by a factor of around two when passivated by SiO_2 and SiN_x as compared to their flat counterparts [CC06, Ker02, MDR+98]. This increase in surface recombination can be attributed on one hand to the geometrical increase in the surface area by a factor of 1.73 as compared to a planar surface (Fig. 4.34), and on the other hand to the poorer passivation of SiO_2 and SiN_x on <111> oriented c-Si as compared to <100> oriented c-Si [RD79, SSES96] (pyramidal facets are <111> oriented). Additionally, in the case of SiO_2, the volume expansion associated with oxidation generates large elastic stresses within the silicon substrate when oxidizing sharp surface features that can even lead to defect creation within the c-Si bulk region [CC06]. Contrariwise, on the single flat <111> oriented c-Si wafer passivated by i a-Si:H, we measured the highest lifetimes out of all a-Si:H/c-Si heterointerface passivation sample structures (Fig. 4.2), and the i a-Si:H passivation of the <111> oriented pyramidal facets should thus be excellent. But in the case of a-Si:H/c-Si heterointerface passivation, the precise texture morphology has shown to be crucial for the passivation quality [AKR+08]. 4.34.

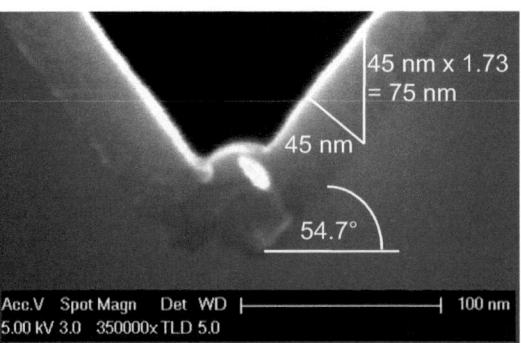

Figure 4.34: *Measurement of a thin-film layer's thickness on textured c-Si by means of a SEM micrograph. To deposit a 45 nm thick layer normal to the c-Si pyramid facets' surfaces, the equivalent thickness of 75 nm on flat c-Si has to be grown, because our a-Si:H grows in the depletion regime.*

4.7. Influence of the texture morphology

As an amorphous silicon thin-film lab, we did not yet dispose of in-house capabilities for texturing. Therefore the first textured wafers were bought thermally oxidized and FGA annealed from the Fraunhofer ISE [ise], ready to be used after a simple HF-dip. For comparison purposes we intended to deposit the same i a-Si:H layer thickness as on flat c-Si. That means we intended to grow the i a-Si:H passivation layer 45 nm thick normal to the c-Si pyramid facets' surfaces as shown on the SEM micrograph (Sec. 2.1.2.2) in Fig.

The pyramidal structure is formed by means of KOH-etching, having a high selectivity of <100> over <111> planes, and therefore, the pyramid angle is given by the c-Si's crystal planes. Thus, because our standard i a-Si:H layer grows in the depletion regime, the a-Si:H deposition time has to be increased by a factor 1.73, i.e., we need to grow the equivalent thickness of 75 nm on flat c-Si. The only way to measure the thickness of an i a-Si:H layer deposited on a textured surface is by SEM or, for even thinner layers, by TEM micrographs. Our method for measuring the thickness of flat layers by alpha-step (Sec. 2.1.1.1) or ellipsometry does not work on a textured surface. As verified with the SEM micrograph shown in Fig. 4.34, by multiplying the deposition time needed to grow 45 nm i a-Si:H on flat c-Si by a factor of 1.73, we effectively grow 45 nm thick i a-Si:H normal to the c-Si pyramidal facets' surfaces. When cleaving such i a-Si:H passivated textured wafers for SEM observations, a change in the fracture surface appears in the pyramidal valley manifesting itself in the appearance of a "crack" irrespective of the perfect cleavage of the c-Si substrate, as shown in Fig. 4.35(a). It is only in imperfect pyramidal valleys, having an increased curvature radius, that homogeneous fracture planes are observed on the c-Si substrate and on the a-Si:H layer, see Fig. 4.35(b). The "crack" that indicates an inhomogeneity in the the density of the a-Si:H layer growth in the pyramidal valleys is only visible after cleavage of the textured c-Si substrate and is not observed when mechanically or chemo-mechanically polishing the same sample for SEM observation. Contrariwise, after the same polishing preparation, a density contrast is visible by TEM in the pyramidal valley throughout the whole i a-Si:H layer grown under increased hydrogen dilution, see again Fig. 4.32(b). Thus, we conclude that in the pyramidal valleys, the growth of the standard i a-Si:H layer is more homogeneous in terms of density than the growth of the a-Si:H layer under increased hydrogen dilution with its initial epitaxy.

4.7. Influence of the texture morphology

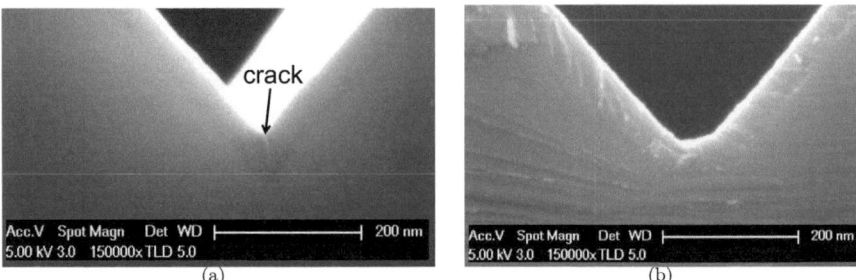

Figure 4.35: *SEM micrograph of i a-Si:H growth in pyramidal valleys of textured c-Si: a) linear inhomogeneity in fracture surface in a-Si:H located at sharp pyramidal valleys on perfectly cleaved c-Si surface and b) same fracture surface of i a-Si:H in pyramidal valleys with increased curvature radius. Note that such "cracks" are not visible in sharp pyramidal valleys of the same sample prepared by polishing for SEM observation.*

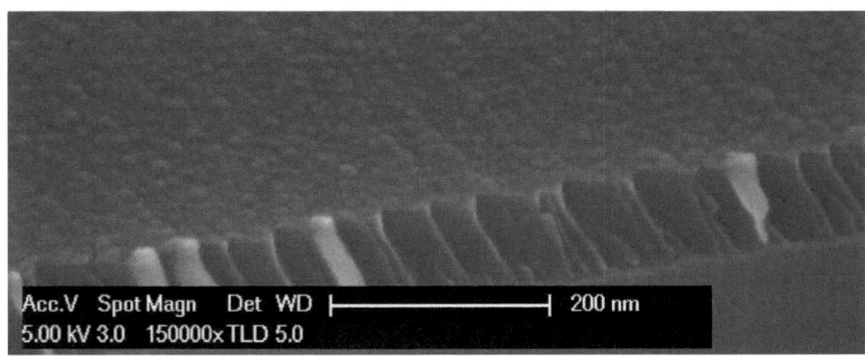

Figure 4.36: *Fracture surface of i a-Si:H cleaved with a different fracture plane from that used in Fig. 4.35 showing the columnar structure and nanoroughness of the grown layer.*

Fig. 4.36 shows that the fracture surface of i a-Si:H changes when cleaved with a different fracture plane from that used in Fig. 4.35. Probably this comes from a columnar structure of the a-Si:H layer that is also the source of the nanometrical surface roughness observed in Fig. 4.36 with lenses of 5 to 30 nm in diameter.

4.7. Influence of the texture morphology

4.7.1 Intrinsic a-Si:H passivation

Passivation of the ISE-textured 1 Ωcm n-type doped wafers with such 45 nm thick i a-Si:H layers results in the $\tau_{eff,m}(\Delta n)$- and $S_{eff,m}(\Delta n)$-curves as displayed simultaneously in Fig. 4.37. With a maximal measured lifetime of 0.35 ms and a minimum $S_{eff,m}$ of 26 cm/s, the passivation quality is significantly lower than on flat c-Si (compare to Figs. 4.2 and 4.3) and would thus result in a maximal open-circuit voltage of 680 mV, thus well below the targeted minimum of 700 mV.

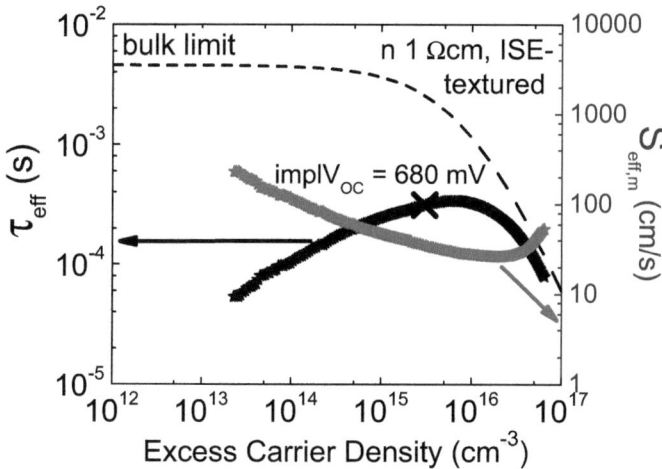

Figure 4.37: *Passivation of ISE-textured 1 Ωcm n-type doped c-Si with 45 nm i a-Si:H. Simultaneous representation of the measured effective lifetime in comparison to the bulk c-Si lifetime limit (black stars, dashed line, lefthand scale), and the effective surface recombination velocity (gray stars, righthand scale). The cross indicates the upper 1-sun open-circuit voltage limit of a solar cell with this surface passivation.*

The wafer's pyramidal surface is made up of nested pyramids and untextured surfaces or surfaces populated with very small pyramids (Fig. 4.38). While such pyramids work well for standard c-Si solar cell fabrication (i.e. with diffused emitters and subsequent SiN_x passivation relating rather on field-effect passivation than on interface defect reduction), they do not permit us to explore the limits of the high passivation performance of our i a-Si:H layers and will limit the open-circuit voltage of finished HJ solar cells.

4.7. Influence of the texture morphology

Figure 4.38: *SEM micrograph of the ISE-textured pyramidal surface showing nested pyramids and untextured surfaces or surfaces populated with very small pyramids.*

In the following, the HMI Germany (now "Helmholtz-Zentrum Berlin für Materialien und Energie") [hmi] kindly provided us some optimally preconditioned [ARK$^+$08] textured c-Si wafers of 1 Ωcm p and n doping type. Preconditioning mainly consists of chemical pyramid facet polishing followed by an RCA clean [Ker90] leaving behind a thin protective oxide, i.e. wafers are ready for use after a HF-dip (Sec. 2.2.1).

Passivation of these wafers with the same 45 nm thick i a-Si:H layers is excellent, as shown in Fig. 4.39(a). Very high maximal lifetime values of 1 ms on n- respectively 0.5 ms on p-type c-Si result in implied open-circuit voltages of over 720 mV on both doping types. The corresponding $S_{eff,m}(\Delta n)$-curves are shown in Fig. 4.39(b), where $S_{eff,m}(\Delta n)$-curves of best flat n- and p-type doped c-Si passivated by i a-Si:H from Fig. 4.3 are shown again for comparison purposes. As argued in Sec. 4.6, we compare $S_{eff,m}$ values at $\Delta n = 5 \times 10^{15}$ cm^{-3} before the strong decrease of $S_{eff,m}$, providing from uncertainties in the bulk lifetime calculation. On the even

4.7. Influence of the texture morphology

more highly n-type doped, textured c-Si the same low $S_{eff,m,min}$ of 5 cm/s as on the 2.8 Ωcm resistive flat c-Si wafer is measured. On p-type c-Si, the textured 1 Ωcm $S_{eff,m,min}$ of 13 cm/s compares to $S_{eff,m,min} = 6$ cm/s on the flat 2.5 Ωcm c-Si wafer. The higher $S_{eff,m,min}$ on 1 Ωcm p-type c-Si could also be related to an overestimated extrinsic bulk lifetime (set here to Yablonovitch's upper limit of $\tau_{extr} = 37$ ms [YAC+86], as a value for the actual τ_{extr} is missing (Sec. 3.2)), because such boron doping levels are observed to yield lower extrinsic c-Si bulk lifetimes in general. Taking into account that the developed surface area is almost double on textured c-Si, interface recombination is thus lower by a factor of almost 2 as compared to flat c-Si suggesting a perfectly textured surface. On one hand, this could be due to the RCA-cleaning before the a-Si:H layer deposition (remember that our flat c-Si wafers are just taken out of the box and HF-dipped). On the other hand, <111> oriented c-Si surfaces are better passivated by i a-Si:H than <100> oriented ones as a minimal surface recombination velocity of as low as 1 cm/s is measured on flat <111> oriented c-Si (Fig. 4.3). The HMI's wafers are thus excellently preconditioned in view of interface defect density reduction by i a-Si:H layer growth. But this wet chemical surface preconditioning is lengthy and, for the moment, the equipment necessary to perform it is lacking at IMT.

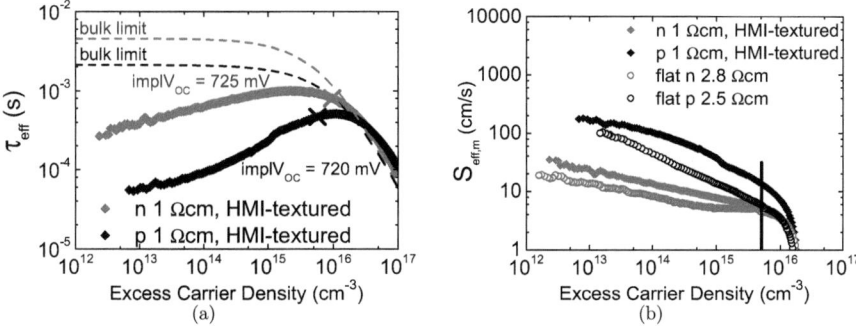

Figure 4.39: *Passivation of optimally preconditioned HMI-textured 1 Ωcm n- and p-type c-Si with 45 nm i a-Si:H. a) Measured effective lifetimes implying V_{OC}s over 720 mV on both substrates and b) effective surface recombination velocities (full dots) compared to the ones of slightly more lightly doped flat c-Si substrates (open dots).*

The success of the HMI-preconditioning is supposed to mainly rely on eliminating the pyramid facets nanoroughness [AKR+08]. Nonetheless, we

4.7. Influence of the texture morphology

suspect that the increased interface defect density observed on the i a-Si:H passivated ISE-textured c-Si is related to the badly defined pyramidal valleys observed in Fig. 4.38. By decreasing the density of poorly defined pyramidal valleys and homogenizing the large pyramidal structure, fewer pyramidal valleys would be found on the same projected area (flat surface), and interface recombination should thus be decreased. To verify this hypothesis, Solarworld [sol] kindly textured us our as-sawn 1.25 Ωcm n-type CZ c-Si wafers with pyramid base lengths ranging from 2 µm to 20 µm average. As shown by the reflectance and transmission spectra in Fig. 4.40, an increased pyramid size does not deteriorate geometrical light-trapping (reflection has the same low value) and the width of only 180 µm for the large pyramidally textured wafer does not increase optical transmission.

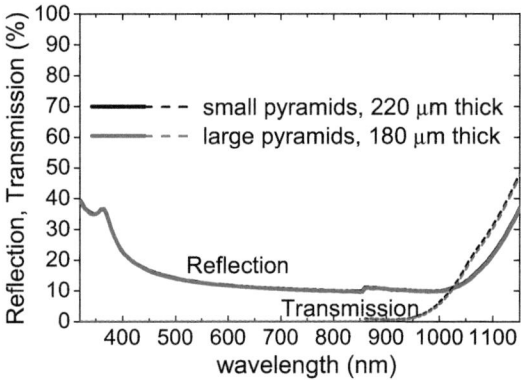

Figure 4.40: *Optical reflection and transmission spectra of c-Si textured with small and large pyramids. Reflection curves overlap and optical transmission is not increased by the reduced substrate thickness of the wafer textured with large pyramids.*

Without further cleaning after texturing, lifetimes measured on these 45 nm i a-Si:H passivated Solarworld-textured samples are very low as shown in Fig. 4.41. Interestingly, the lifetime measured on the wafer with the largest pyramids is nonetheless higher than the one measured on the smallest sized pyramids. Those wafers could then be cleaned in another lab of our university by a simplified RCA-clean (exact process sequence in Sec. 2.2.2). Figure 4.42 shows the 45 nm thick i a-Si:H passivation performance on these cleaned wafers with varying pyramid sizes. The passivation result of the same i a-Si:H layer on the ISE-textured and the optimally HMI-preconditioned wafers are replicated by smaller symbols for comparison.

4.7. Influence of the texture morphology

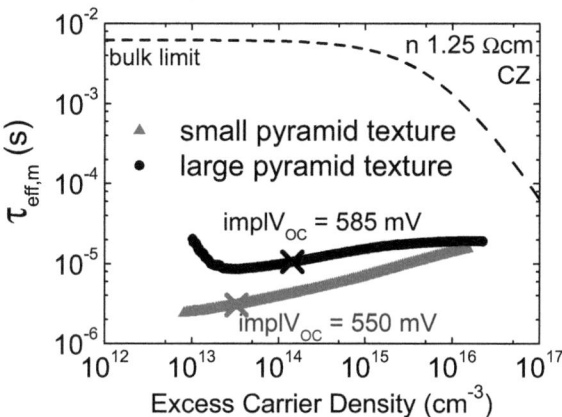

Figure 4.41: *Passivation performance of 45 nm i a-Si:H on as-textured non-cleaned wafers prepared by Solarworld with varying pyramid base length. As a consequence of the low measured lifetimes, low V_{OC}s of final solar cells would result. Note that compared to the other $\tau_{eff}(\Delta n)$-curves in this study, the lifetime scale on this figure had to be extended to lower values in order to be able to display the low measured lifetimes.*

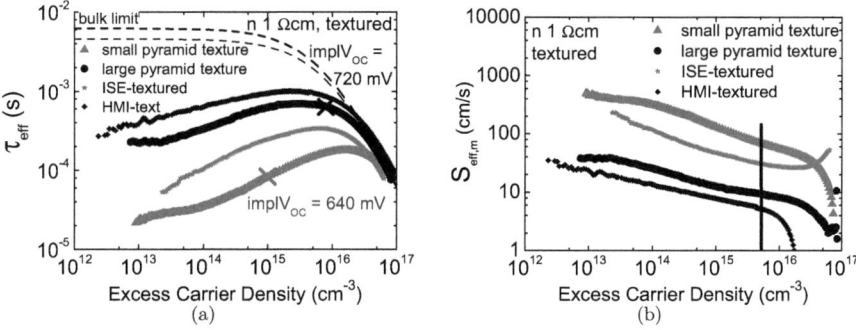

Figure 4.42: *Cleaned Solarworld-textured c-Si of varying pyramid size passivated by 45 nm i a-Si:H. For comparison purposes, our results on ISE- and HMI-textured c-Si wafers passivated with the same layers are indicated with smaller symbols. a) Lifetime measurements and b) effective surface recombination velocities.*

4.7. Influence of the texture morphology

On one hand, as expected, the i a-Si:H passivation of the largest pyramids is very good, yielding maximal lifetimes of 0.7 ms, i.e. $S_{eff,m,min}(\Delta n = 5 \times 10^{15}$ cm$^{-3}) = 10$ cm/s and again an implied V_{OC} notably over 700 mV, i.e. impl$V_{OC} = 720$ mV. On the other hand, the same i a-Si:H passivation yields, on the smallest pyramids, a maximal lifetime of scarcely 0.2 ms, that translates into a $S_{eff,m,min}(\Delta n = 5 \times 10^{15}$ cm$^{-3})$ of 70 cm/s. The interface defect density reduction of i a-Si:H is thus an order of magnitude lower on these smallest pyramids as compared to the largest ones. As a consequence, the open-circuit voltage limit imposed by recombination is 640 mV on the smallest pyramids, i.e. 80 mV lower than on the largest ones. The i a-Si:H layer's lower passivation performance on the small pyramidal textured wafer is possibly related to its high density of pyramidal valleys where the i a-Si:H layer shows growth inhomogeneities (not epitaxy, Fig. 4.35(a) and related comments). The interface defect density increase by almost an order of magnitude ($S_{eff,m,min}$ of 70 cm/s versus 10 cm/s) is in good agreement with the large to small pyramid density ratio of $\frac{20 \times 20}{2 \times 2} = 100$. The high passivation performance on a large pyramidally textured surface is rather homogeneous as shown by ILM lifetime mappings (Sec. 2.1.2.1) performed on such a wafer passivated by 100 nm i a-Si:H in Fig. 4.43.

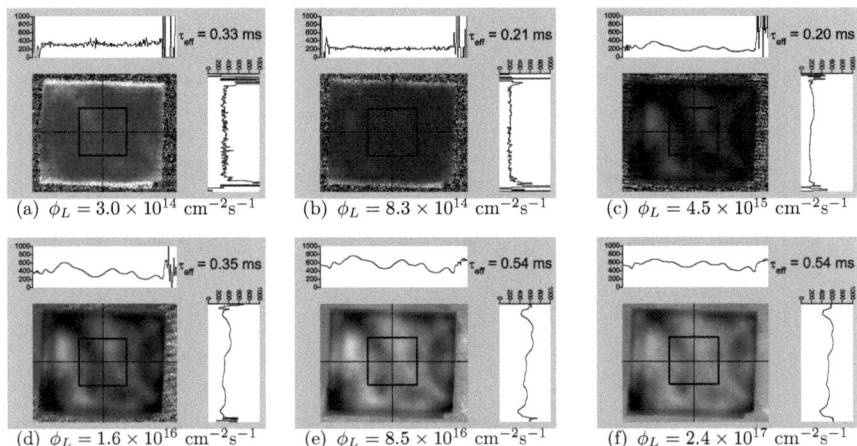

Figure 4.43: *ILM mappings made under varying photon flux densities, increasing from a) to h). The square of 3 cm side length denotes the region over which $\tau_{eff,m}$ is averaged to calculate the $\tau_{eff,ILM}(\Delta n)$-curve in Fig. 4.44.*

4.7. Influence of the texture morphology

The observed small lifetime inhomogeneities result either from the CZ-wafer itself, from the cleaning process or from passivation inhomogeneities due to texture nature variations on the wafer. The comparison with flat c-Si in Fig. 4.8 indicates that i a-Si:H layers very homogeneously passivate the substrates, i.e. that an inhomogeneous deposition in our VHF-reactor is not the cause of the variations observed in Fig. 4.43. Again, there is excellent agreement between Sinton lifetime measurements and ILM lifetime measurements averaged over a square corresponding about to the Sinton lifetime tester's sensing coil area, as shown in Fig. 4.44.

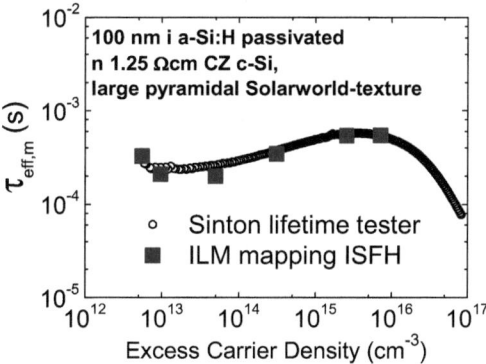

Figure 4.44: *Excellent accordance of Sinton lifetime tester measurements with ILM mappings of 100 nm i a-Si:H passivated n 1.25 Ωcm c-Si, textured with large pyramids by Solarworld. The measurement area of the Sinton coil (diameter of 3.5 cm) corresponds about to the square (3 cm side length) over which $\tau_{eff,ILM}(\Delta n)$ is calculated from Fig. 4.43.*

4.7.2 Emitter and BSF layer stack passivation

In analogy to flat c-Si (Sec. 4.5), both the emitter and BSF layer stacks are separately symmetrically grown on textured c-Si, to test their suitability for Si heterojunction solar cell formation. We verify that the growth of the same layer thickness perpendicular to the pyramidal surface as compared to the flat c-Si substrate's surface needs a deposition time multiplication by a factor of 1.73, as effectively suggested by the observation of directional growth and the geometrical considerations in Fig. 4.34. While the same thickness perpendicular to the microscopic c-Si surface is supposed to yield the same passivation, most of the light is nonetheless incident normally to the whole c-Si substrate. But thanks to the very high refractive index

4.7. Influence of the texture morphology

of Si in the highly absorbed short-wavelength range ($n_r = 5$ for a-Si:H and $n_r = 4.5$ for μc-Si:H in the highly absorbing short-wavelength range), light is refracted almost normal to the pyramidal facets, see Fig. 2.16, and emitter absorption is thus not increased as compared to flat c-Si. In view of not hindering carrier transport through the emitter layer stack in complete Si HJ cells (Sec. 5.2), we choose nonetheless a deposition time increase by only a factor of 1.4 (instead of 1.73) for the device-layer equivalent symmetrical textured c-Si passivation. Figure 4.45 shows the passivation performance of the symmetrically grown i/p emitter and three different i/n BSF layer stacks on the ISE-textured n-type c-Si wafer compared to the standard thicker i a-Si:H layer passivation taken from Fig. 4.37.

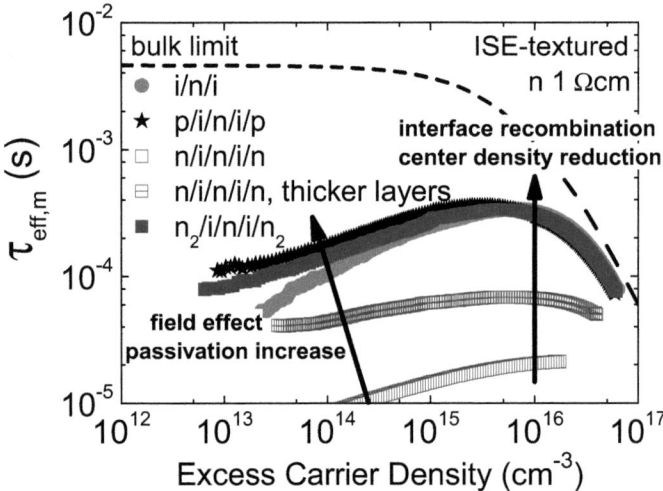

Figure 4.45: *Passivation performance of symmetrically grown emitter (i/p) and three different BSF (i/n) layer stacks on ISE-textured n-type doped c-Si (1 Ωcm) compared to i a-Si:H passivation. The standard n-layer performing well on flat c-Si limits V_{OC} to 630 mV on textured c-Si. New process parameters for the n_2-layer are finally found for a good BSF layer stack passivation performance.*

In the same manner as on flat n-type c-Si (Fig. 4.25), compared to the thicker i a-Si:H single layer, the emitter layer stack adds field-effect passivation (also Fig. 4.27 for illustration purposes), without greatly increasing the interface defect density. Contrariwise, the standard n a-Si:H/μc-Si:H transition layer grown for BSF formation greatly increases the inter-

4.7. Influence of the texture morphology

face defect density. Because a-Si:H is transparent to the long-wavelength light reaching the BSF layer stack (see absorption coefficients/penetration depths in Si in Fig. 2.2), for a better protection of the a-Si:H/c-Si interface against ion bombardment during the n a-Si:H/µc-Si:H transition layer growth, the thickness of the back's i a-Si:H passivation layer perpendicularly to the pyramidal surface is increased to its equivalent thickness on flat c-Si, i.e. the deposition time of the back i a-Si:H passivation layer is multiplied by a factor of 1.73 instead of 1.4 as compared to flat c-Si. Also the n a-Si:H/µc-Si:H transition layer's growth time is prolonged in order to permit its specific microstructure's development that ensures an efficient doping and thus an efficient field-effect passivation. Although, as desired, the interface defect density is reduced and field-effect passivation enhanced, a solar cell's V_{OC} incorporating this BSF layer stack would be limited to 630 mV. Because further fine-tuning of the deposition parameters of this BSF stack, well performing on flat wafers, does not lead to better performances, a CO_2 plasma treatment has been intercaled in the n a-Si:H/µc-Si:H deposition sequence (see Sec. 2.3.4). Optimization of these oxidizing plasma parameters finally yields a n_2 BSF layer. In the case of a-Si:H solar cells, it is observed that a hydrogen plasma on an i a-Si:H layer such as performed when growing the standard i/n BSF layer stack, increases its optical bandgap but also its dangling bond density [Per01]. This new n_2-layer's passivation performance in stack with i a-Si:H as BSF is equivalent to the passivation performance of the i/p emitter layer stack, but its lower conductivity as compared to the standard n-layer directly reflects in a lower field-effect passivation (again Fig. 4.45). Because of the weaker n_2-based BSF, FF-losses have to be expected in the Si HJ solar cell application. Contrariwise, on flat c-Si, the performance of this n_2-layer included in solar cells' BSF layer stacks turned out to be significantly lower than the standard n-layer. The question, why different n-layer parameters are optimal in terms of interface passivation on flat, respectively textured c-Si, remains open and will be partially answered by the TEM micrograph analysis in Sec. 5.5.

From the observed epitaxial growth of a highly hydrogen diluted a-Si:H layer in pyramidal valleys of textured c-Si and the additionally observed inhomogeneities under cleavage even of the standard i a-Si:H layer, the question about the nature of the c-Si/ultrathin a-Si:H/µc-Si:H crystallographic interfaces in such pyramidal valleys arises. As shown in the case of the flat Si HJ solar cell, the abrupt growth of the doped a-Si:H/µc-Si:H transition layer on the standard i a-Si:H layer (which is growing abruptly amorphous on c-Si (Fig. 4.28)) is verified by TEM observations (Fig. 5.13).

4.7. Influence of the texture morphology

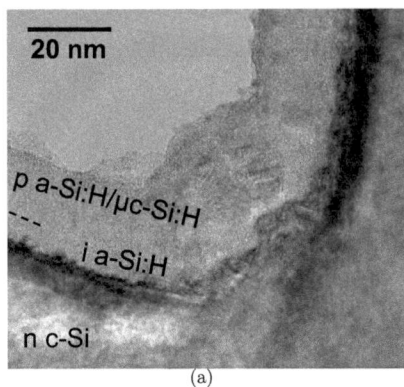
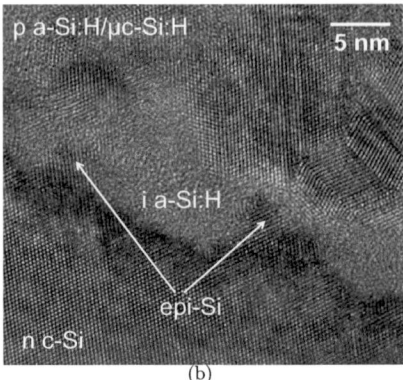

Figure 4.46: *a) TEM and b) HR-TEM micrographs of a textured Si HJ solar cell emitter layer stack consisting of n c-Si / i a-Si:H / p a-Si:H/µc-Si:H. In comparison with a flat Si HJ solar cell emitter shown in Fig. 5.13, one can observe epitaxy of the a-Si:H based passivation layers located at the bottom of the pyramid valleys. Such local epitaxial regions were identified as unsatisfactory passivated spots at the a-Si:H/c-Si interface (Figs. 4.32 and 4.33).*

TEM observations of the i/p emitter layer stack's growth on the large pyramidal Solarworld-textured CZ c-Si wafer show the same abrupt interface between the three materials as on flat c-Si, but epitaxially grown regions can be identified in the pyramidal valleys as shown in Fig. 4.46. Such local epitaxial regions have been identified as unsatisfactory passivated spots at the a-Si:H/c-Si interface (passivation decrease in Fig. 4.33(b) coming from the epitaxial structure in the pyramidal valley shown in Fig. 4.32). Nonetheless, the passivation quality decrease, evaluated by means of the lifetime measurement in Fig. 4.47 and compared to the thicker intrinsic standard layer's passivation (taken from Fig. 4.33(a)), is minor thanks to the compensation of the interface defect density increase by additional field-effect passivation. This is in contrast to the marked passivation decrease observed with smaller epitaxial structures in the case of i a-Si:H passivation relying only on the interface defect decrease for passivation (Fig. 4.33(b)). In addition to this effect, and compared to the ISE-textured c-Si, the large pyramidal Solarworld-textured c-Si was observed to be free of flats (untextured surfaces) and showed a very limited presence of mini-pyramids under SEM (compare the SEM micrographs in Fig. 4.38 and Fig. 5.22(a), where the latter is taken on medium sized Solarworld-textured c-Si).

4.7. Influence of the texture morphology

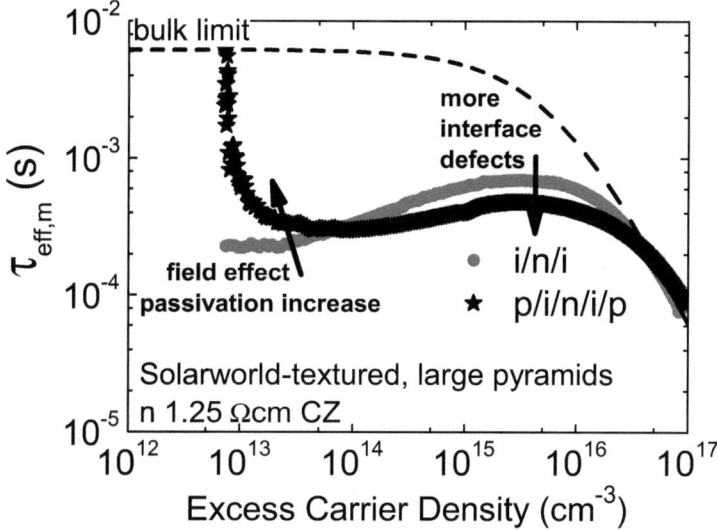

Figure 4.47: *Solarworld-textured n-type CZ c-Si passivation with the symmetrically grown i/p emitter layer stacks, compared to thicker i a-Si:H passivation layers. The local epitaxial regions identified on this sample (Fig. 4.46) deteriorate the passivation performance much less than do the ones observed within a thicker highly H_2-diluted i a-Si:H layer (Fig. 4.33(a)) because of the additional field-effect passivation mechanism of the emitter layer stack.*

The device-quality i a-Si:H standard layer of this i/p stack was replaced by the non-diluted one, whose exact process parameters are given in Tab. 2.2, under the hypothesis that a non-hydrogen diluted i a-Si:H layer would reduce the observed local epitaxial growth by preventing the silicon atoms' rearrangement on the c-Si surface thanks to their reduced surface mobility. But in fact, the reduced surface mobility of the depositing Si atoms grown without hydrogen dilution leads to a poorer conformal substrate coverage of the growing layer and we observe even local epitaxial growth where two pyramid walls intercept with a wide angle. The omission of hydrogen-dilution from the i a-Si:H layer's growth conditions increases epitaxial growth in the somewhat <100> oriented pyramidal valleys as compared to flat c-Si where an increase in hydrogen dilution finally leads to epitaxial growth. This favors a picture where stress/strain rather than surface orientation is the reason for the occurrence of local epitaxial

4.7. Influence of the texture morphology

growth in the pyramidal valleys, because the lower surface atom mobility during deposition leads to higher stress. The passivation performance of the non-diluted i a-Si:H / p a-Si:H/µc-Si:H layer stack is shown in Fig. 4.48. Compared to the slightly diluted case, a more epitaxially "short-circuited" i a-Si:H passivation layer leads, as expected, to a slight interface defect density increase accompanied by an enhanced field-effect passivation. This observation is in good accordance with the lower V_{OC}s (more interface defects) but high FFs (high field-effect passivation) observed on Si HJ solar cells fabricated without i a-Si:H passivation layers [KCA+07].

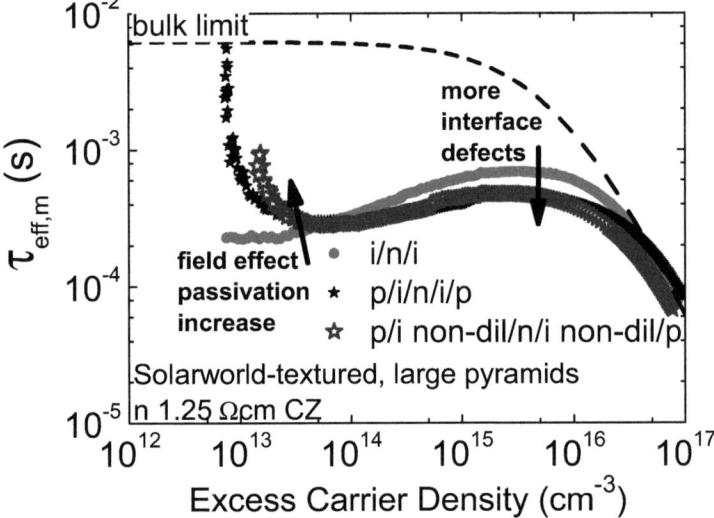

Figure 4.48: *Performance of the pure silane i a-Si:H / p a-Si:H/µc-Si:H emitter layer stack passivating symmetrically n-type CZ c-Si textured with large pyramids (fabricated by Solarworld), compared to the standard i/p emitter layer stack grown under device-quality i a-Si:H layer dilution. Because of increased local epitaxy in the non-diluted i a-Si:H layer case, interface defect density is slightly increased but field-effect passivation is enhanced.*

4.8 Limits imposed on V_{OC} and FF by interface recombination: choice of the optimal c-Si doping type and level for Si HJ solar cell fabrication

Recombination losses set an upper limit on finished solar cell's performances. In this section, we address the question of the best suited doping type and level for Si HJ solar cell fabrication from a material specifics interface recombination properties point of view. From the simultaneous measurement of the wafer's photoconductivity (coil) and the light intensity decay (photodetector), an implied V_{OC} as a function of the illumination level can be calculated (Sec. 2.1.2.1). These data can be further analyzed to indicate the JV-curve that a completed HJ solar cell containing this interface passivation would have, if its performances would be only limited by a-Si:H/c-Si interface recombination. An implied current is determined for each open-circuit voltage from Eq. 2.20. 1-sun short-circuit current densities of 35 mA/cm^2 and 39 mA/cm^2 are assumed for excellent planar respectively textured Si HJ solar cells. From the resulting implied JV-curve, the limit imposed on FF by recombination can be extracted.

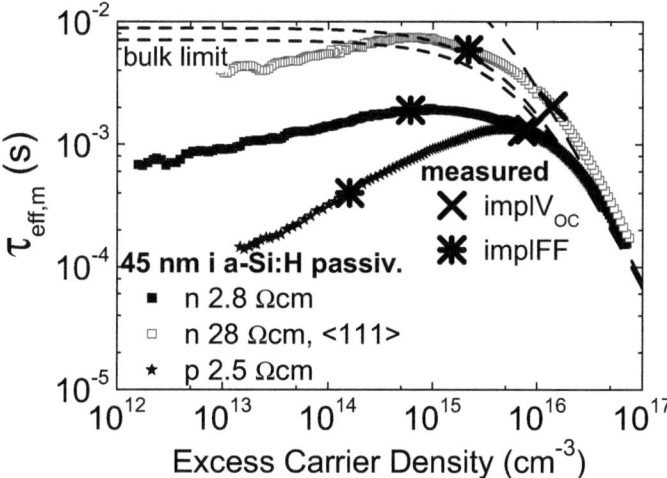

Figure 4.49: *Selection of $\tau_{eff,m}(\Delta n)$-curves from the i a-Si:H passivations in Fig. 4.2 with implied V_{OC}s and FFs indicated at their corresponding injection level. Extracted values are summarized in Tab. 4.4.*

4.8. Limits imposed on V_{OC} and FF by interface recombination: choice of the optimal c-Si doping type and level for Si HJ solar cell fabrication

Figure 4.49 shows a selection of three $\tau_{eff,m}(\Delta n)$-curves from the i a-Si:H passivations in Fig. 4.2, where the implied V_{OC} and FF are indicated at their corresponding injection level, and their values are given in Tab. 4.4.

	implV_{OC}	implFF
n 2.8 Ωcm	715 mV	80.5%
n <111> 28 Ωcm	735 mV	82.5%
p 2.5 Ωcm	715 mV	78.0%

Table 4.4: *Upper limits imposed on V_{OC} and FF by recombination (implV_{OC} and implFF [%]) extracted from the injection level dependent lifetime measurement shown in Fig. 4.49.*

Modeling a-Si:H/c-Si interface recombination with the a-Si:H/c-Si amphoteric interface recombination parameters (Sec. 3.5.3) is in good agreement with experimental results. From the calculated $\tau_{eff,c}(\Delta n)$-curve an implied current-voltage curve, i.e. a JV-curve without series resistance effects and possible carrier transport problems, can be calculated as well, as shown in this section. 1-sun illumination corresponds to a photon flux ϕ_L of 2.75×10^{17} cm^{-2}s^{-1} impinging on the c-Si surface. The resulting generation rate in a c-Si substrate is then given from Eq. 2.9 by the illuminated surface's optical transmission factor F and the c-Si substrate thickness W, assuming all photons entering the c-Si substrate are absorbed. The 1-sun generation rate of a given passivated sample is related to the 1-sun lifetime in the steady-state condition through the injection level, i.e. Eq. 2.10:

$$\tau_{1-\text{sun}} = \frac{\Delta n}{G_L(1 \text{ sun})}. \qquad (4.3)$$

The $\tau_{eff,c}(\Delta n)$-curve calculated for this specific c-Si wafer with the a-Si:H interface passivation properties (neglecting the optical absorption losses within the passivation layer and transmission losses throughout the whole sample) intersects this $\tau_{1-\text{sun}}(\Delta n)$-curve at Δn corresponding to 1-sun illumination for this wafer with this passivation, i.e. $\tau_{eff,c}(\Delta n(1 \text{ sun})) = \tau_{1-\text{sun}}(\Delta n(1 \text{ sun}))$. Therefore the implied 1-sun V_{OC} is calculated from Eq. 2.13 by

$$\text{impl}V_{OC}(1 \text{ sun}) = \frac{kT}{q} ln[\frac{(n_0 + \Delta n(1 \text{ sun}))(p_0 + \Delta p(1 \text{ sun}))}{n_i^2}]. \qquad (4.4)$$

The total recombination current J_{rec} is given by Eq. 3.25, $J_{rec} = q \times W \times R_{tot}$, where the total recombination rate R_{tot} [cm^{-3}s^{-1}] is the sum of

4.8. Limits imposed on V_{OC} and FF by interface recombination: choice of the optimal c-Si doping type and level for Si HJ solar cell fabrication

surface, Auger, radiative and extrinsic bulk c-Si recombination (Sec. 3.2). The current-voltage characteristic of a solar cell limited by recombination can then be expressed in analogy to Eq. 2.20 (Sec. 2.1.2.6) as

$$J_{impl} = J_{SC}(1 - \frac{J_{rec}}{J_{rec}(\Delta n(1 \text{ sun}))}). \quad (4.5)$$

In view of choosing the optimal doping type and level for Si HJ solar cell formation, implied V_{OC}s and FFs are calculated with the a-Si:H/c-Si interface recombination specific parameters without additional field-effect passivation for two different interface dangling bond densities and for flat and textured c-Si, both 200 μm thick. Table 4.5 summarizes implied V_{OC}/FF pairs as well as the corresponding efficiencies when varying the wafer doping type and level for the case of low and very low interface dangling bond densities for flat c-Si. Table 4.6 shows the same for textured c-Si.

$N_s = 10^{10}$cm^{-2}	impl$V_{OC}/FF/Eff$			$N_s = 3 \times 10^9$cm^{-2}	impl$V_{OC}/FF/Eff$		
n 1 Ωcm	685 mV	84.0%	20.1%	n 1 Ωcm	715 mV	85.5%	21.4%
n 5 Ωcm	675 mV	82.0%	19.4%	n 5 Ωcm	715 mV	84.5%	21.1%
n 28 Ωcm	675 mV	80.0%	18.9%	n 28 Ωcm	720 mV	85.0%	21.4%
p 0.5 Ωcm	650 mV	83.5%	19.0%	p 0.5 Ωcm	685 mV	84.0%	20.1%
p 3 Ωcm	640 mV	78.5%	17.6%	p 3 Ωcm	715 mV	84.5%	21.1%
p 5 Ωcm	655 mV	76.0%	17.4%	p 5 Ωcm	715 mV	84.5%	21.1%

Table 4.5: *Flat c-Si: implied V_{OC} and FF for different doping types and levels and resulting efficiency assuming $J_{SC} = 35$ mA/cm^2, calculated with the amphoteric interface recombination model for low and very low interface dangling bond density N_s (no field-effect passivation).*

$N_s = 10^{10}$cm^{-2}	impl$V_{OC}/FF/Eff$			$N_s = 3 \times 10^9$cm^{-2}	impl$V_{OC}/FF/Eff$		
n 1 Ωcm	700 mV	84.5%	23.1%	n 1 Ωcm	725 mV	85.5%	24.2%
n 5 Ωcm	695 mV	82.0%	22.2%	n 5 Ωcm	730 mV	85.5%	24.3%
n 28 Ωcm	690 mV	81.0%	21.8%	n 28 Ωcm	730 mV	85.5%	24.3%
p 0.5 Ωcm	660 mV	83.5%	21.5%	p 0.5 Ωcm	700 mV	84.5%	23.1%
p 3 Ωcm	670 mV	78.5%	20.5%	p 3 Ωcm	725 mV	85.0%	24.0%
p 5 Ωcm	685 mV	77.5%	20.7%	p 5 Ωcm	725 mV	85.5%	24.2%

Table 4.6: *Textured c-Si: implied V_{OC} and FF for different doping types and levels and resulting efficiency assuming $J_{SC} = 39$ mA/cm^2, calculated with the amphoteric interface recombination model for low and very low interface dangling bond density N_s (no field-effect passivation).*

Fig. 4.50 shows the calculated underlying $\tau_{eff,c}(\Delta n)$-curves indicating the injection level corresponding to implV_{OC} and implFF for the four cases

4.8. Limits imposed on V_{OC} and FF by interface recombination: choice of the optimal c-Si doping type and level for Si HJ solar cell fabrication

(low - very low interface dangling bond density / flat - textured c-Si) at the example of Fig. 4.50(a) 1 Ωcm n-type and Fig. 4.50(b) 5 Ωcm p-type doped c-Si.

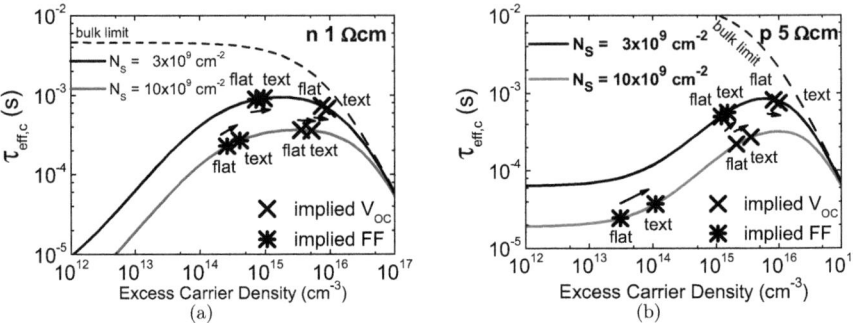

Figure 4.50: *Calculated $\tau_{eff,c}(\Delta n)$-curves of a-Si:H passivating a) 1 Ωcm n-type and b) 5 Ωcm p-type c-Si. The a-Si:H/c-Si interface recombination calculation permits the extraction of the upper limits set on V_{OC} and FF by recombination, as listed in Tabs. 4.5 and 4.6. The injection level corresponding to V_{OC} and FF conditions is indicated for low and very low interface dangling bond density for flat and textured c-Si respectively.*

Figure 4.51 graphically summarizes the wafer doping type and level dependent Si HJ solar cell efficiency limits for low/very low interface dangling bond density N_s and flat/textured c-Si calculated with our DB interface recombination model. First of all, we see that 1-5 Ωcm n-type doped c-Si results in the best solar cells. For more highly doping, V_{OC} (Tab. 4.5) becomes limited by bulk c-Si recombination, namely Auger recombination. The use of p-type doped c-Si results in lower performing solar cells. The more the interface is defective, the more pronounced the difference between n- and p-type is. Solar cells based on lightly doped wafers suffer from a severe fill factor limitation (Tab. 4.5) as long as the interface dangling bond density is not sufficiently decreased. These predictions are in good accordance with our experimental observations. When using textured c-Si, reflection losses are minimized and photogeneration in the c-Si wafer is therefore higher. The higher generation rate increases the 1-sun injection level and therefore shifts it into lower interface recombination regions (Fig. 4.50), and enhances the implied V_{OC} and FF, as listed in Tab. 4.6. In view of process stability, i.e. the reduced sensitivity to an increased interface

4.8. Limits imposed on V_{OC} and FF by interface recombination: choice of the optimal c-Si doping type and level for Si HJ solar cell fabrication

dangling bond density, 1 Ωcm n-type doped c-Si is the first choice for complete high efficiency device processing, as is most easily recognized from Fig. 4.51. If one would even more reduce the interface defect density, the solar cell efficiency would finally be limited by intrinsic c-Si recombination only (supposed $\tau_{defect} = \infty$) to 29% [GKV98].

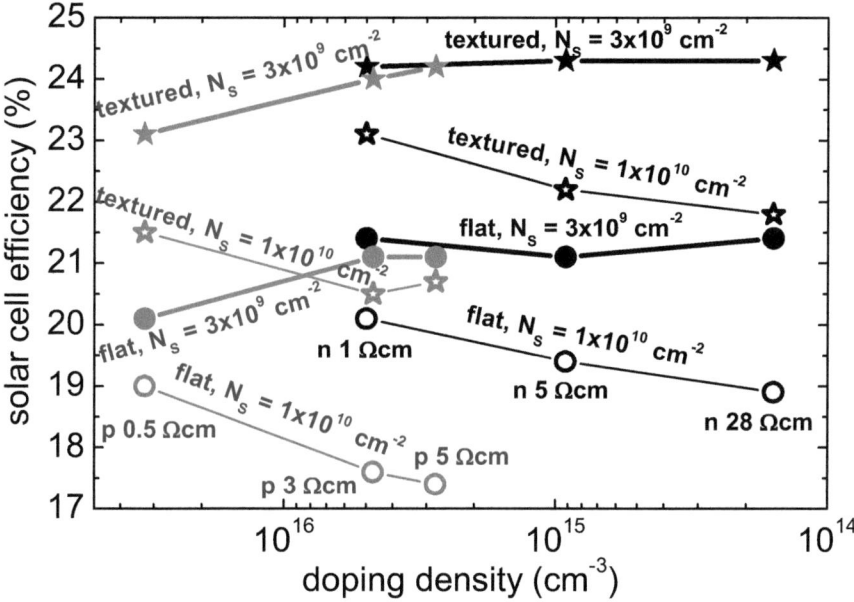

Figure 4.51: *Graphical summary of efficiency limits imposed by interface recombination as calculated with our amphoteric interface recombination model for low/very low interface DB density N_s and flat/textured c-Si (assuming $J_{SC} = 35 \ mA/cm^2$ for flat and $J_{SC} = 39 \ mA/cm^2$ for textured c-Si).*

In practice, series resistances resulting from thin-film Si layers and TCO resistances as well as the effects of blocking junctions dominate the value of FF. As the real V_{OC} is not affected by series resistances, implV_{OC} is in general reached in final solar cells given that the ITO deposition is sufficiently soft and does not lead to an increased defect density at the a-Si:H/c-Si interface.

4.9 Conclusions on optimized a-Si:H/c-Si interface passivation

In conclusion, intrinsic a-Si:H acts as a high performance passivation layer on all kind of c-Si substrates. This simple, low temperature passivation process does not need to fear comparison to other high performing passivation schemes such as SiO_2 and SiN_x. The i a-Si:H's excellent passivation quality is identified in this work to be related to a marked decrease of the interface dangling bond density. This conclusion is reached by modeling interface recombination by means of amphoteric dangling bonds, as observed in bulk a-Si:H. The a-Si:H/c-Si interface recombination limit imposed on V_{OC} is pushed to record values well over 700 mV on all substrates studied experimentally, that include FZ and CZ of various doping.

Field-effect passivation can be added by microdoping i a-Si:H layers or fixing their outer surface potential by an overlying doped a-Si:H layer. In contrast to SiO_2 and SiN_x, the lowest effective surface recombination velocity (1 cm/s corresponding to a lifetime of 7.5 ms) is measured on <111> oriented c-Si. Because textured, complete high performance solar cells rely on the passivation of <111>-oriented pyramidal facets, this is an important finding.

The abruptness of the crystallographic nature of the a-Si:H/c-Si heterointerface plays an important role in determining the interface passivation quality. Abrupt interfaces are observed when passivating flat c-Si by VHF-PECVD grown thin-film Si layers (and layer stacks), provided that epitaxial growth is avoided by choosing reasonable PECVD process parameters. This optimal crystallographic interface is much more challenging to obtain in textured c-Si substrates. Lifetime measurements on differently textured c-Si substrates emphasize the importance of the texture's size and quality. With the appropriate surface preconditioning, the same high implV_{OC}s of well over 700 mV, as on flat c-Si, are reached by i a-Si:H passivation. TEM micrographs of i a-Si:H plus doped a-Si:H/μc-Si:H layer stacks, as used as emitter and BSF in Si HJ solar cells, reveal the presence of epitaxial growth at the bottom of the pyramidal valleys. These features are identified here as the cause of the poorer performances of i a-Si:H passivation on textured c-Si. Therefore, contrariwise to flat c-Si, the substrate's crystalline nature is not lost through the i a-Si:H interlayer for the subsequent growth of the doped a-Si:H/μc-Si:H emitter and BSF layer. In contrast to the flat doped a-Si:H/μc-Si:H transition layers' preliminary development on glass, layer characterization tools do not work any longer on textured c-Si. Therefore, it is only by means of lifetime measurements

4.9. Conclusions on optimized a-Si:H/c-Si interface passivation

that the doped a-Si:H/µc-Si:H layers' growth conditions can be adapted to the textured surface's nature and optimized passivation layer stacks are developed. Finally, using fully amorphous doped Si layers (still in stack with i a-Si:H) instead of a-Si:H/µc-Si:H transition layers for emitter and BSF formation in Si HJ solar cells, this detrimental epitaxial growth could possibly be prevented.

Chapter 5

Amorphous/crystalline silicon heterojunction solar cells

First of all, the carrier transport through the interfaces in Si HJ solar cells is briefly addressed in this chapter (Sec. 5.2). Lifetime measurement guided Si HJ solar cell optimization by device diagnostics is then presented, where some redundancy to Chap. 4 (a-Si:H/c-Si interface passivation) is included in this new context of Si HJ solar cells (Sec. 5.3). From there on, the achievement of well performing flat Si HJ solar cells is straight-forward (Sec. 5.4). On textured c-Si, TEM micrographs greatly assist lifetime measurements by identifying local epitaxial growth in pyramidal valleys as the main concern for textured Si HJ solar cell optimization (Sec. 5.5). This chapter is supposed to be self-consistent.

5.1 Introduction

Silicon heterojunction solar cells made by Sanyo, called HIT (heterojunction with intrinsic thin-layer) [WTS+91], achieve record efficiencies of 22.3% with $V_{OC} = 725$ mV, $J_{SC} = 38.9$ mA/cm^2 and $FF = 79.1\%$ on a total area of 100.5 cm^2 as confirmed by AIST [TYT+09]. The fabrication of a-Si:H/c-Si heterojunction solar cells is of great interest for several reasons:

- A fabrication process at low temperature (200 °C), which reduces the wafer breakage (in particular when Al layers are used) and the energy necessary to invest in the fabrication process.

5.1. Introduction

- As ultra-low surface recombination is achieved and the low fabrication temperature reduces material breakage, thinner wafers can be used, resulting in a strong reduction in the consumption of silicon (with a long-term potential of 4 g/Wp against $10-11$ g/Wp of standard c-Si solar cells, including material losses).

- The possibility to achieve ultra high solar cell efficiency $> 20\%$.

- A better temperature coefficient than standard c-Si solar cells, i.e. a reduction from $-0.5\%/°C$ to $-0.25\%/°C$.

- As the solar cell fabrication process is now based on thin-film silicon, it offers the prospect of applying the cost-effective industrial deposition techniques used for thin Si layers.

Therefore, Si HJ solar cells combine the best of crystalline silicon solar cells on one hand and amorphous Si solar cells on the other hand, illustrated in Fig. 5.1. That is why this kind of ultra-high efficient solar cell is one of the most promising candidates to reach grid parity in the segment of highly efficient c-Si cells.

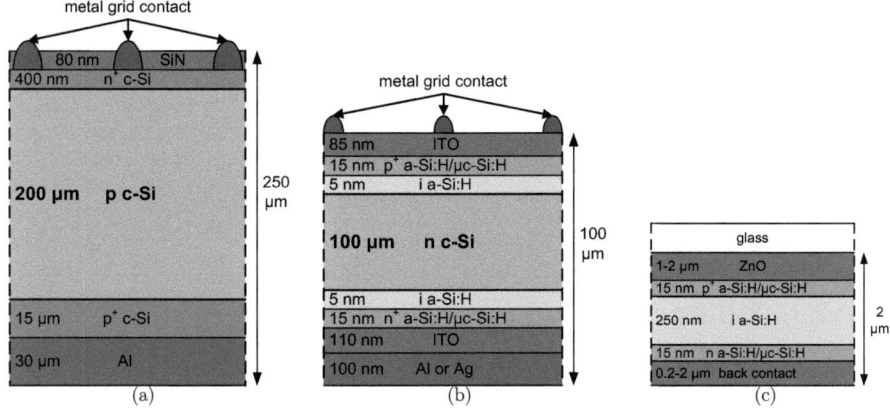

Figure 5.1: *a) Conventional crystalline Si solar cell, b) Si HJ solar cell and c) amorphous Si solar cell.*

The interpretation of lifetime measurements is a powerful tool for the characterization of the performances of a-Si:H layers and layer stacks with respect to their passivation properties in Si HJ solar cells. However, the final judgement about an emitter and back surface thin-film Si layer stack can only be made once it has been incorporated into a complete solar cell. Besides current transport issues due to the a-Si:H/c-Si HJ, the contact between the Si-based stack and the front TCO (see Fig. 5.1(b)) can further limit the device performance. In practice, ITO does not make an ideal ohmic contact to the p-type thin-film Si layer, and therefore its work function is crucial. However, thanks to lifetime measurements on heterostructure test samples, we dispose of a tool for single process step analysis that allows a rapid high-efficiency Si HJ solar cell development. This shows up to be especially useful on textured c-Si where most layer characterization tools no longer work.

5.2 Carrier transport in a-Si:H/c-Si heterojunction solar cells

The upper efficiency limit imposed on a specific type of wafer by i a-Si:H/c-Si interface recombination depends only on the interface dangling bond density. The open-circuit voltage in complete Si HJ solar cells is still solely determined by the interface dangling bond density. However, carrier transport of the a-Si:H/µc-Si:H layer stack and thus the solar cell FF is dominated by i) the i a-Si:H layer thickness, ii) its dangling bond density, iii) band offsets to c-Si, iv) the doping level of the p- and n-type a-Si:H/µc-Si:H layers and v) last but not least the latter's contact to the TCO.

Taguchi *et al.* [TMT08] recently confirmed that the forward current in Si HJ solar cells is determined by diffusion currents at higher bias voltages (> 0.4 V) and by tunneling currents at lower bias voltages (< 0.4 V). While the open-circuit voltage increases with the i a-Si:H passivation layer thickness due to an improved interface dangling bond density reduction, the Si HJ solar cell's FF shows the inverse trend [TMT08]: the thinner the i-layer, the easier for the photogenerated minority carriers in the c-Si to cross the barrier at the interface to the a-Si:H by a tunneling process as shown by the band diagram in Fig. 5.2 [vCSR98]. Contrariwise, a less doped emitter layer inhibits tunneling [vCRR[+]98] and in more defective i a-Si:H layers, carriers recombine before being collected [DBL[+]08]. The collection of photogenerated carriers is critically dependent on the band offsets, i.e. mainly on the valence band offset ΔE_V for n-type c-Si HJ solar cells and mainly on the conduction band offset ΔE_C for p-type c-Si HJ solar cells (the definitions of these variables are given in Fig. 5.4). Van Cleef

5.2. Carrier transport in a-Si:H/c-Si heterojunction solar cells

et al. [vCRR+98] demonstrate by simulation that too high a valence band offset hinders the transport of holes to a p^+ a-SiC:H emitter layer in case of n-type c-Si HJ solar cells. There are big discrepancies in literature about these band offset values ranging from much higher valence than conduction band offsets to the inverse [RÖ3, DL07]. These band offsets are supposed to strongly depend on the deposition conditions and on the thin-film Si layer's doping [BOD+07].

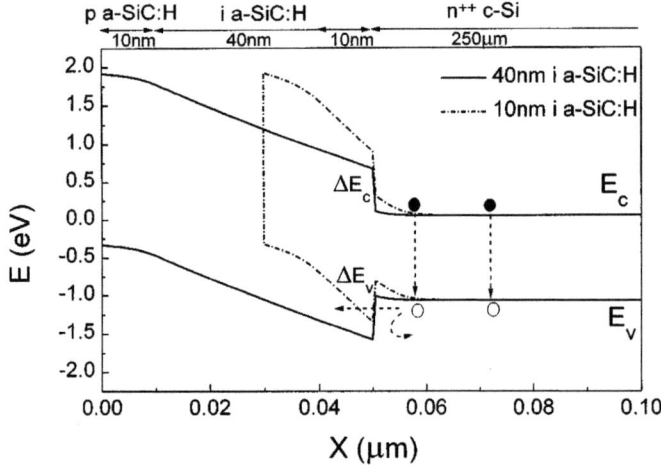

Figure 5.2: *Calculated band diagram of a heterojunction emitter on n-type c-Si with AMPS at zero bias voltage, taken from [vCSR98]: the thinner the i a-Si:H layer, the easier for the holes to cross the barrier at the interface to a-Si:H by a tunneling process.*

To achieve high V_{OC}s, tunneling has to be inhibited, whereas for high FFs a sufficient amount of tunneling is needed. Thus, best solar cell efficiencies are reached by fine-tuning this amount of tunneling by means of bandgap and band offset engineering, while always maintaining a minimal i a-Si:H layer dangling bond density in i a-Si:H layers as well as a maximized band bending, either by very highly doped thin-film Si emitter and BSF layers or by well chosen TCO work functions. However, we achieve good results on both wafer types, see implied V_{OC}s of i a-Si:H passivation in Tab. 4.2 and solar cells in Sec. 5.4, and ascribe our more severe FF limitation on p-type c-Si to the interface recombination properties (Sec. 4.8) rather than to a much higher valence band offset.

5.2. Carrier transport in a-Si:H/c-Si heterojunction solar cells

For the illustration of collection problems, Fig. 5.3 shows a typical S-shaped 1-sun JV-curve resulting here from a contaminated PECVD chamber. Lifetime measurements on the co-deposited precursor (process path 4 in Fig. 5.5) yield the same implied V_{OC} as measured by the SunsV_{OC}-curve on the finished cell, as always when the appropriate ITO deposition parameters are used (Sec. 2.4.1). Although the SunsV_{OC} implied FF is already severely limited by recombination, only the 1-sun JV-curve measurement exhibits an S-shape and a very low FF, indicating the serious collection problems of this solar cell (Fig. 5.3).

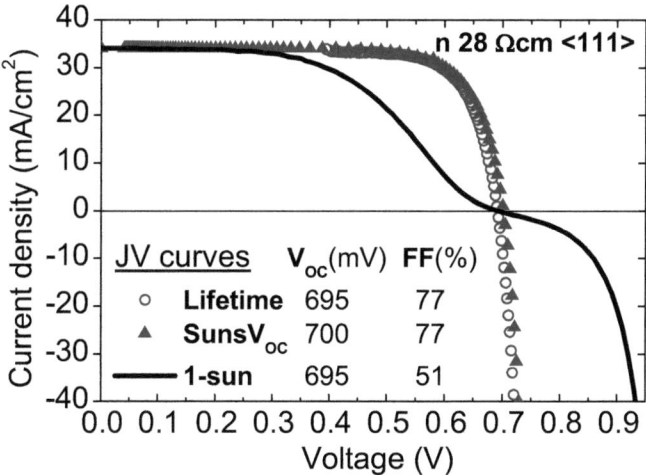

Figure 5.3: *S-shaped 1-sun JV-curve resulting from a contamination of the VHF-PECVD chamber. Lifetime measurements on the precursor (process path 4 in Fig. 5.5) correctly predict the final 1-sun V_{OC}. This cell's severe collection problems do not result from the injection level-dependance of recombination, but are probably dominated by recombination within the i a-Si:H layer.*

Last but not least, the ITO work function that is the energy difference between the vacuum level and the ITO's Fermi level, can deteriorate the carrier collection [Ien04]. The ITO's work function ϕ_{ITO} is usually smaller than $\phi_{p\,a-Si:H}$ (Fig. 5.4 and Sec. 2.4.1 for variable definitions), and therefore hinders hole tunneling from the c-Si into the thin p a-Si:H layer when brought into contact. One way to circumvent this problem is to use a microcrystalline p-type emitter layer that can be much more highly doped,

5.2. Carrier transport in a-Si:H/c-Si heterojunction solar cells

forming an ohmic contact with the ITO. Contrariwise, the n a-Si:H/μc-Si:H layer could probably be replaced by a purely amorphous n a-Si:H layer (n a-Si:H can be much more highly doped than p a-Si:H) improving process stability. However, without increasing our ITO's work function, no high FF will be reached with a p a-Si:H instead of a p a-Si:H/μc-Si:H layer before ITO deposition.

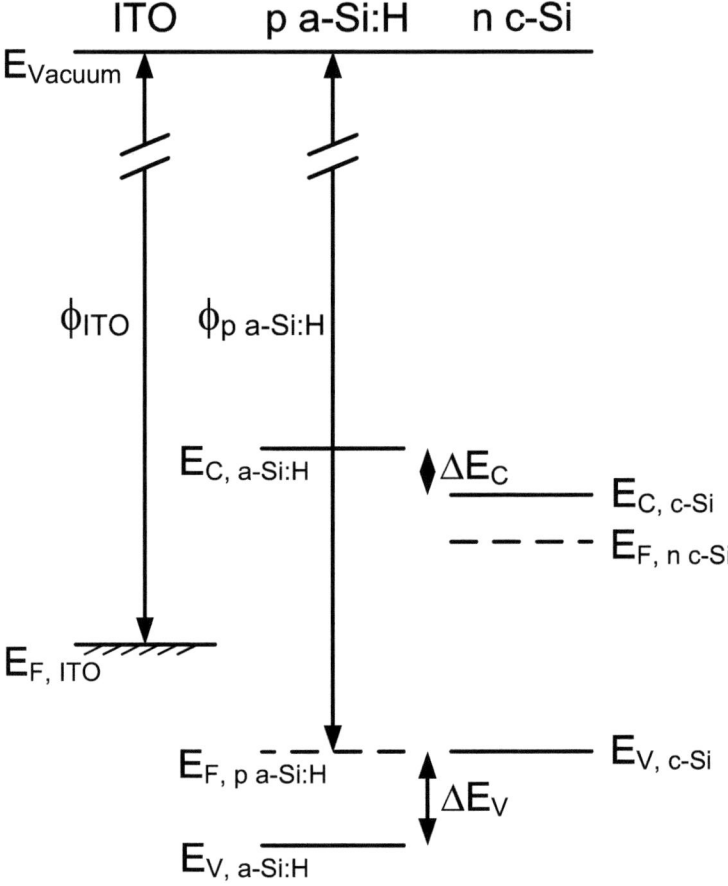

Figure 5.4: *Definitions of valence and conduction band offsets ΔE_V and ΔE_C [eV] as well as ITO work function ϕ_{ITO} [V] and p a-Si:H Fermi level $E_{F,p\,a-Si:H}$ [eV] with respect to vacuum $\phi_{p\,a-Si:H}$ [V].*

5.3. Lifetime measurements as a guide for solar cell optimization

5.3 Lifetime measurements as a guide for solar cell optimization

The complete fabrication of a Si HJ solar cell requires several steps, as indicated in Fig. 5.5. The possible erroneous process step responsible for a final device with poor performance is hard to detect. However, with the Sinton lifetime tester, we dispose of a tool for single process step analysis. Figure 5.5 shows the process flow of the various a-Si:H/c-Si lifetime test structures, paths 1 − 4, used for a fast device diagnostic of the complete solar cell fabrication in path 5, where the corresponding description of the detailed process flow of complete a-Si:H/c-Si HJ solar cell formation is given in Sec. 2.5.2. Path 0 is useful to check very roughly the bulk lifetime and surface cleanliness of a c-Si wafer as well as the diluted HF's state. For example, on an as-sawn or an as-textured wafer, a simple HF-dip will yield no passivation due to the highly defective c-Si surface, respectively the c-Si's surface contamination. However, lifetime after HF-dip rapidly decreases due to fast reoxidation (Fig. 2.12(b)), and, therefore, the measured lifetime will critically depend on the time before making the measurement.

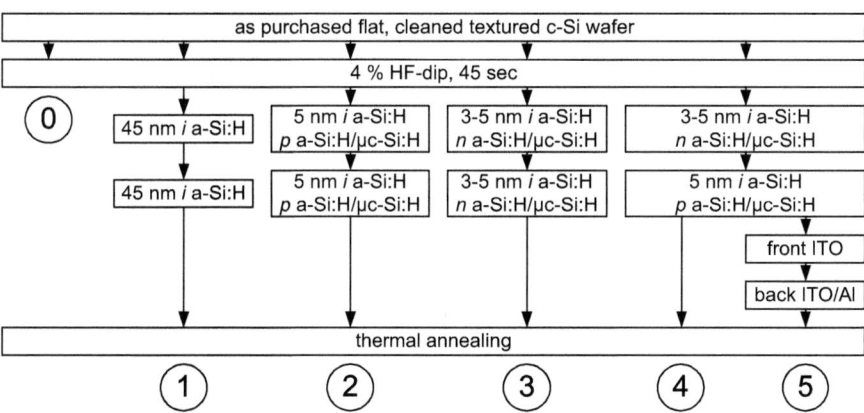

Figure 5.5: *Process flow of various a-Si:H/c-Si lifetime test structures (paths 1 − 4) allowing for a fast device diagnostic of complete solar cell fabrication (path 5) by means of a single process step analysis capability.*

Sec. 4.8 discusses how the injection level-dependance of the effective lifetime can be transformed into an illumination dependent implied V_{OC}-curve and furthermore into an implied JV-curve. This allows a useful

171

5.3. Lifetime measurements as a guide for solar cell optimization

determination of the limits on V_{OC} and FF solely imposed by recombination. Conversely, together with the measurement of the final solar cell's JV-curve, injection level dependent lifetime measurements allow for device diagnostics after single process steps. This allows the identification of which step was defective whenever the predicted solar cell upper performance limits are not reached. The efficiency of the diagnostic based on lifetime test structures is illustrated further by the example of a solar cell having a lower V_{OC} than expected. Figure 5.6 shows the JV-curve (line) of a complete solar cell resulting from path 5 (Fig. 5.5), and the implied JV-curve (open dots) of its co-deposited precursor from path 4. The comparison of these two curves permits a diagnostic of the impact of the ITO deposition on the a-Si:H/c-Si HJ solar cell precursors, e.g. an interface dangling bond density increase by too heavy a surface bombardment during ITO deposition.

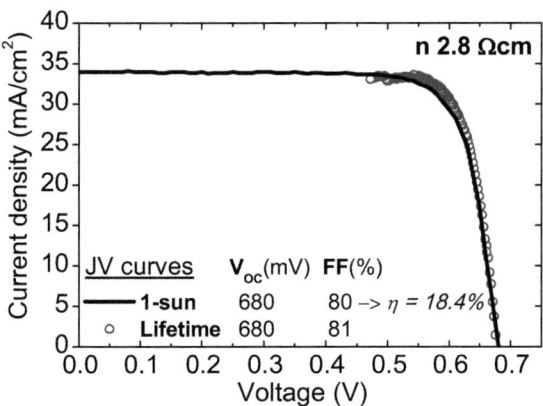

Figure 5.6: *1-sun JV-curve of a complete solar cell having a low V_{OC} compared to the JV-curve implied by lifetime measurements (Sec. 4.8) on the same sample before ITO deposition. The accordance between the two curves identifies interface recombination losses as the source for this V_{OC} limitation. The BSF layer stack is finally identified as the cause of the low V_{OC} of the complete solar cell (Fig. 5.7).*

From Fig. 5.6 it is obvious that the final ITO deposition does not increase recombination and introduces only minor series resistance losses. The solar cell's V_{OC} is thus limited by interface recombination. Lifetime measurements on the test structures whose process flows are given in Fig. 5.5, path 1 to 3, permit us to identify the source of these interface recombi-

5.3. Lifetime measurements as a guide for solar cell optimization

nation losses. After having concluded that the 45 nm thick i a-Si:H passivation layer quality is highly reproducible, the purity of the HF-solution and the sufficient cleanliness of the VHF-PECVD deposition chamber can be checked by fabricating such lifetime test structures. The $\tau_{eff,m}(\Delta n)$-curve resulting from path 1 shown in Fig. 5.7 is very high, implying a V_{OC} of 715 mV. In consequence, the unexpectedly lower solar cell V_{OC} of 680 mV in Fig. 5.6 is neither caused by insufficient surface cleaning or contaminated HF nor by the i a-Si:H layer's quality. Suspecting that the p-layer deposition deteriorates the overall lifetime as observed by others [DWB06, GRB+05], the passivation performance of the i/p stack is tested (path 2 in Fig. 5.5).

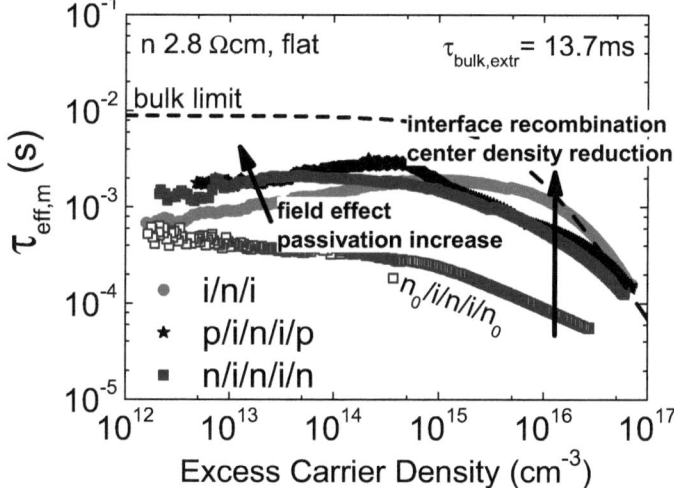

Figure 5.7: *Single process step analysis procedure by means of lifetime test structures identifying the BSF layer stack ($n_0/i/n/i/n_0$) as the source of the low V_{OC} of the device presented in Fig. 5.6. Improving this BSF layer stack's growth conditions yields the solar cell properties shown in Fig. 5.8.*

But as shown in Fig. 5.7, the deposition of the thinner i a-Si:H layer and the further p-type doped a-Si:H/μc-Si:H transition layer growth only slightly increase the a-Si:H/c-Si interface dangling bond density compared to the thicker i a-Si:H layer. Because this overlying p a-Si:H/μc-Si:H layer adds field-effect passivation, it is sufficiently doped and well suited as an emitter in Si HJ solar cells. Surprisingly, it is the deposition of the remaining i/n stack (denoted n_0 in Fig. 5.7) that is, in this case, the cause of the low V_{OC} of the solar cell in Fig. 5.6. The lifetime measurements

5.3. Lifetime measurements as a guide for solar cell optimization

on the test structure resulting from process path 3 in Fig. 5.5 show that this n-layer deposition increases the interface dangling bond density while being nonetheless able to establish field-effect passivation.

This additional field-effect passivation identified within the i/p and the i/n layer stack leads to the high FF of the solar cell made out of this emitter and BSF layer stacks, proving its excellent collection and contact properties. Improving the growth condition of the n-layer results then in the same low interface dangling bond density as achieved by the i/p stack without loosing in field-effect passivation. The newly fabricated solar cell precursor (path 4 in Fig. 5.5) containing the now optimized i/n-stack (filled square symbols in Fig. 5.7) yields a promising implied V_{OC} of 710 mV (Fig. 5.8). But as seen from the simultaneous representation of the $SunsV_{OC}$ JV-curve (full triangles) in Fig. 5.8, the ITO deposition is detrimental as it causes increased recombination and consequently diminishes the finished solar cell's V_{OC}. Note that this cell's ITO was deposited immediately after a target exchange. This study led us to re-adapt the ITO deposition process parameters resulting in the ITO$_2$ recipe (Sec. 2.4.1). The final comparison to the 1-sun JV-curve (line) indicates appreciable additional "series resistance" losses, see discussion about the current transport in Si HJ solar cells in Sec. 5.2.

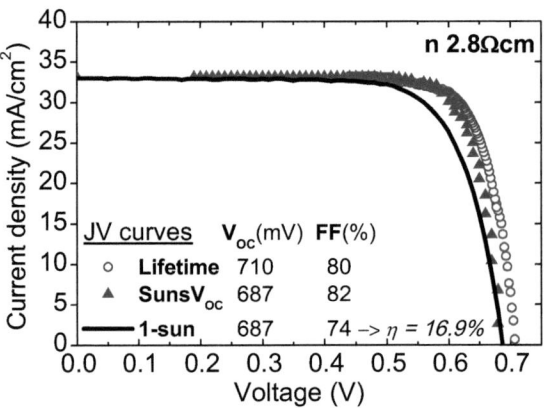

Figure 5.8: *JV-curve implied by lifetime measurements on the solar cell precursor (process path 4 in Fig. 5.5) compared to the JV-curves measured on the finished solar cell. Their comparison allows the identification of this solar cell's 1-sun V_{OC} being limited by an increased interface dangling bond density resulting from the ITO deposition.*

5.4 a-Si:H/c-Si heterojunction solar cells based on flat c-Si

The intrinsic a-Si:H layer improves the open-circuit voltage in final Si HJ solar cells but leads to lower fill factors when deposited too thickly, see Sec. 5.2. This trade-off between a high V_{OC} and a high FF is well illustrated by the JV-curves of our best performing and our highest-V_{OC} Si HJ solar cells shown in Fig. 5.9. Note that the highest efficiencies are achieved on 1 Ωcm n-type doped c-Si whereas the highest V_{OC} is obtained on <111>-oriented c-Si, also n-type, but more lightly doped. All three cells use ITO$_1$ (Sec. 2.4.1).

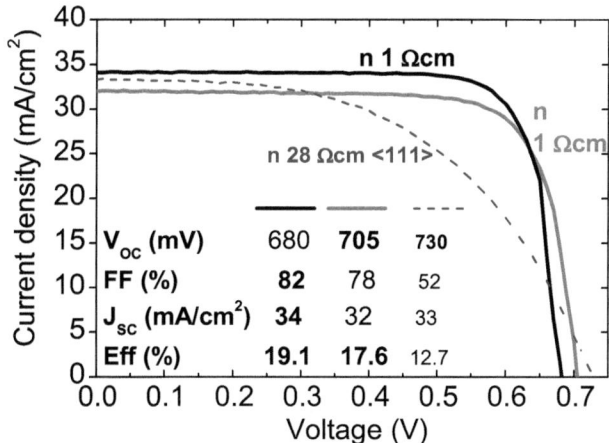

Figure 5.9: JV-curves of our highest-efficient and highest-V_{OC} flat a-Si:H/c-Si heterojunction solar cells. Note that J_{SC} is determined by EQE measurements (Sec. 2.1.2.5).

It should be noted that our solar cell efficiency measurements are subject to uncertainties because of multiple measurement difficulties. In particular, ITO is directly contacted by measurement probes, and because of the small size of our cells (4.5 mm × 4.5 mm) edge effects are important (structuring into single cells, lateral collection).

On one hand, a high efficiency of 19.1% due to an excellent FF and J_{SC} is achieved, while on the other hand an excellent V_{OC} of 730 mV is obtained because of a measured lifetime of 1.7 ms at 1 sun (on this precursor lifetime exceeds 10 ms at an injection level of 5×10^{14} cm^{-3}!). Also on p-

5.4. a-Si:H/c-Si heterojunction solar cells based on flat c-Si

type c-Si, 0.5 Ωcm doped, we were able to achieve a high V_{OC} of 690 mV when depositing the same Si thin-film layers, whose process parameters are given in Sec. 2.3.2 and 2.3.4, i.e. the i/n layer stack as emitter and the i/p layer stack as BSF. However, this p-type c-Si HJ solar cell had a lower efficiency of 16.3% mainly due to its lower FF of 74% partially limited by the i a-Si:H layers' passivation properties on p-type c-Si, see Sec. 4.8. Other groups report high FFs also on p-type c-Si but without including an i a-Si:H passivation layer for emitter and BSF formation, and thus lower V_{OC}s [KCA$^+$07]. Due to the poorly defined size of our small cells, the short-circuit current density J_{SC} is evaluated from the EQE measurement, see Sec. 2.1.2.5. Figure 5.10 shows the EQE of a flat Si HJ solar cell together with the IQE, calculated by taking into account the optical reflection losses.

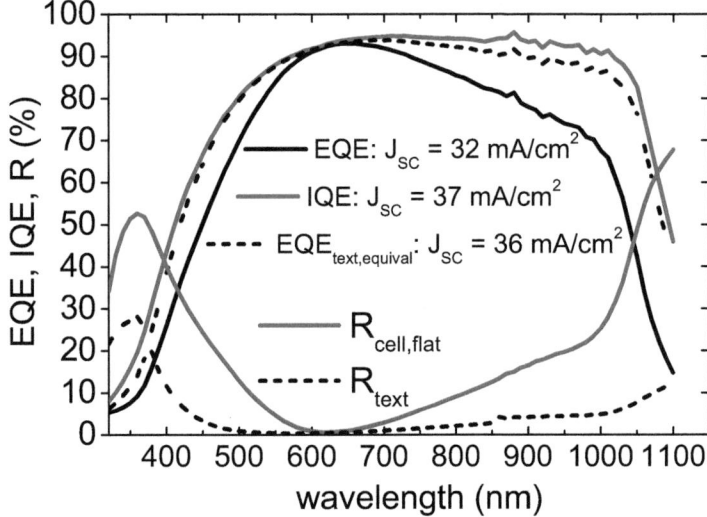

Figure 5.10: *EQE of a Si HJ solar cell, such as used for J_{SC} calculation, compared to the IQE calculated from this cell's reflection spectrum. On flat c-Si, 5 mA/cm^2 of short-circuit current density are lost by reflection, mainly on the single layer front antireflection coating (ITO), neglecting long-wavelength absorption losses in the back metal contact. Using textured instead of flat c-Si greatly reduces reflection losses, mainly by front surface double reflection, see Sec. 2.4.1. In this specific solar cell, one could gain 4 mA/cm^2 of J_{SC}.*

As observed from the big discrepancies of IQE and EQE in Fig. 5.10,

5.4. a-Si:H/c-Si heterojunction solar cells based on flat c-Si

reflection losses are high due to the flat substrate nature in combination with a single layer antireflection coating with a low refractive index at $\lambda = 1000$ nm in addition, see Sec. 2.4.1, in particular Fig. 2.15. Knowing the spectral photon flux of sunlight on earth, the short-circuit current density that is reached by a given QE curve can be calculated by Eq. 2.19, Sec. 2.1.2.5. Without reflection losses (neglecting long-wavelength absorption losses in the back metal contact), this specific solar cell would reach $J_{SC,IQE} = 37$ mA/cm^2 instead of the effectively measured J_{SC} of 32 mA/cm^2. 5 mA/cm^2 are thus lost by reflection. Using textured c-Si (reflection curve in Fig. 5.10 taken from Fig. 2.17), the reflection losses of this specific solar cell could be reduced to 1 mA/cm^2 compared to the upper boundary of 37 mA/cm^2 set by absorption in ITO and thin-film Si based layers and by recombination at the a-Si:H/c-Si interfaces. A net current gain of 4 mA/cm^2 would thus be expected by using textured instead of flat c-Si on this specific solar cell. The highest efficient Si HJ solar cell of Sanyo reaches $J_{SC} = 39$ mA/cm^2 (including shadowing by the contact grid). Compared to our highest J_{SC} of 34 mA/cm^2 reached on flat c-Si, there is a current gain potential of up to 15% when using textured instead of flat c-Si. In comparison to the 42 mA/cm^2 measured on the highest-efficient c-Si cells (PERL) [ZWGF98], 8 mA/cm^2 are lost by our best flat cell. In Si HJ solar cells, losses due to absorption without photogeneration in the ITO and the emitter thin-film layer stack add to reflection losses (Sec. 2.1.2.5). Figure 5.11 shows the same cell as Fig. 5.10 with a low $J_{SC} = 32$ mA/cm^2. The ITO layer's free carrier absorption leads to a short-circuit current density loss of 2.5 mA/cm^2 ($J_{SC,IQE^*} - J_{SC,IQE}$). This loss can be reduced by decreasing the ITO layer's free carrier density while simultaneously increasing its free carrier's mobility to keep the required low resistance (Sec. 2.4.1). An additional 2.5 mA/cm^2 are lost by absorption in the emitter layer stack in the worst case ($J_{SC,IQE^{**}} - J_{SC,IQE^*}$), i.e. assuming it to be completely photoelectrically inactive. This value can be reduced by reducing the emitter layer stack's thickness. Compared to the IQE of the highest-efficient c-Si solar cell [ZWGF98], an additional 2 mA/cm^2 of short-circuit current density are lost by front and back surface recombination in this specific Si HJ solar cell ($J_{SC,IQE^{**}} - J_{SC,IQE,PERL}$).

5.4. a-Si:H/c-Si heterojunction solar cells based on flat c-Si

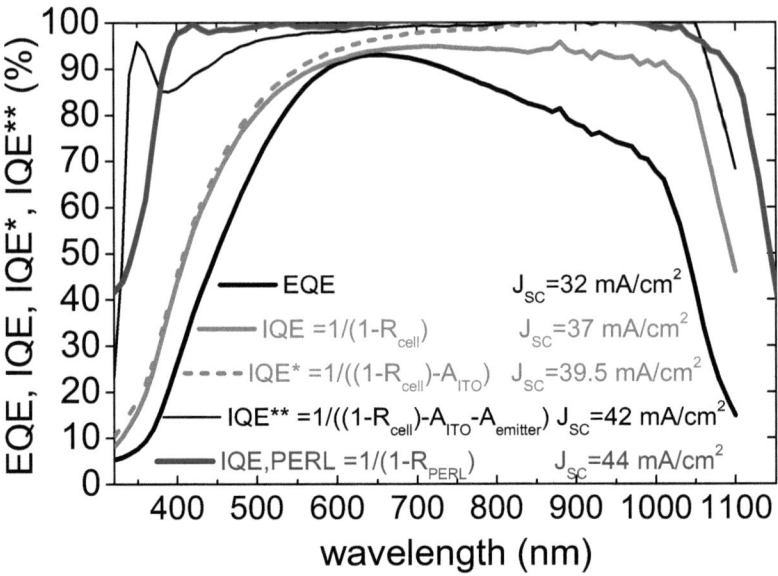

Figure 5.11: *Analysis of short-circuit current density losses and comparison to the highest-efficient c-Si solar cell (PERL) [ZWGF98]. 5 mA/cm² of short-circuit current density are lost by reflection on the flat front surface, an additional 2.5 mA/cm² by ITO absorption and 2.5 mA/cm² more by emitter absorption. The remaining difference of 2 mA/cm² to the highest-efficient c-Si solar cell are lost by interface recombination.*

We thought it would be realistic to attain a HJ solar cell efficiency of 19.5% on flat c-Si by fine-tuning the thin-film Si's deposition parameters with special attention to the interface treatments (CO_2 and H_2 (Sec. 2.3.4)). Unfortunately we became aware of silicon thin-film deposition reproducibility problems. They were identified as mainly due to the single deposition chamber without a plasma confinement box and without gas cleaning configuration (Sec. 2.3.1). A plasma confinement box has proven to greatly help to better define the process parameters. Additionally, depositing intrinsic and doped layers all in the same chamber without gas cleaning capability, leads easily to cross-contamination and necessitates frequent cleaning by hand, where the latter severely decreases the reproducibility. Especially the doped layers for emitter and BSF formation that grow at the edge of the amorphous to microcrystalline Si transition are affected by these plasma process parameter stability problems. Amorphous doped layers would be less sensitive to process parameter instabilities. But our standard ITO layer does not produce a good contact to p-type a-Si:H layers (Sec. 5.2) and our ITO's amelioration in this respect is hindered by the process parameter space limitations in the disposable system (Sec. 2.4.1).

The same high V_{OC}s of well over 700 mV could be reached again. However, now having different probes at our disposal for JV-curve measurements we became aware that positioning probes directly on ITO and not on a metal grid is problematic for correct FF measurements. That is why for the fine-tuning of efficiency optimization the very small area (~ 0.2 cm^2) gridless solar cells should be replaced by at least 1 cm^2 area sized cells with an evaporated contact grid on top.

5.5 Textured a-Si:H/c-Si heterojunction solar cells

The short-circuit current density of flat Si HJ solar cells is estimated to be limited to 35 mA/cm^2 by front reflectance losses, whereas 34 mA/cm^2 are reached by our best flat cell (Sec. 5.4). Compared to Sanyo's record HIT solar cell's short-circuit current density of 39 mA/cm^2, there is a current gain potential of 15% for our best flat Si HJ solar cell. This is achieved by reducing front reflection losses through geometrical light-trapping by means of pyramidally texturing the c-Si's surface (Secs. 2.4.1 and 5.4).

5.5.1 Introduction

In the case of flat c-Si, the emitter and BSF layer stack optimization based on thickness, conductivity and crystallinity measurements on glass and then on lifetime test structures, leads straight-forward to good flat Si HJ solar cell performances. Section 4.7 already highlighted the more challenging passivation of textured c-Si and is intimately related to this section. Exactly the same layer stacks yielding good emitters and BSFs on flat c-Si perform badly on textured c-Si, as shows their JV-curve comparison in Fig. 5.12.

This textured solar cell is based on our first textured wafers (thermally oxidized and FGA annealed) that we bought from the Fraunhofer ISE [ise]. The standard i a-Si:H passivation reveals a V_{OC}-limitation to under 680 mV because of these wafers' specific surface morphology, see Sec. 4.7 and the V_{OC} effectively attained by this textured Si HJ solar cell is with 610 mV

5.5. Textured a-Si:H/c-Si heterojunction solar cells

even markedly below this limit.

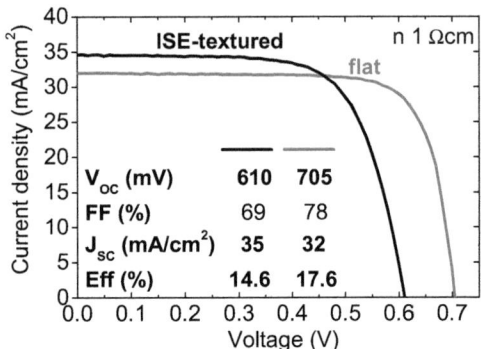

Figure 5.12: *Comparison of a-Si:H/c-Si HJ solar cell performances on flat and textured c-Si being based on the same 1 Ωcm n-type doped wafer and exactly the same process parameters for thin-film Si emitter and BSF layer stack growth. Instead of the expected efficiency gain thanks to a light-trapping enhanced current, the textured c-Si based solar cell's efficiency is lowered as compared to flat c-Si. Thus, a one to one transfer of process parameters from flat to textured c-Si does not give HJ cells with higher performances.*

In fact, neither do we know the respective thicknesses of the grown layers nor do we know if their microstructure and thus their conductivity is as desired. If a layer whose process parameters are chosen to yield amorphous/microcrystalline transition material grows fully amorphous instead, a poorly doped layer results. In contrast, for application on flat c-Si, these layers are easily characterized on glass, see Sec. 2.1.

HR-TEM micrographs prepared by the cleaved corner method, Fig. 5.13(a), as well as by polishing, Fig. 5.13(b), confirm the abrupt nature of the interfaces in our best performing flat a-Si:H/c-Si solar cells consisting of c-Si / i a-Si:H / doped a-Si:H/µc-Si:H layers.

In fact, for purely thin-film Si based solar cell formation, i.e. a-Si:H, µc-Si:H and a-Si:H/µc-Si:H tandem solar cell fabrication, thicker layer growth has been shown to be much more critical on textured than on flat c-Si. Figure 5.14(a) shows the non-conformal growth of a-Si:H on a textured substrate, whereas in Fig. 5.14(b) growth inhomogeneities of µc-Si:H in sharp pinches are observed [PVSB+08]. The latter is shown to deteriorate the µc-Si:H solar cell's performances [PVSB+08].

5.5. Textured a-Si:H/c-Si heterojunction solar cells

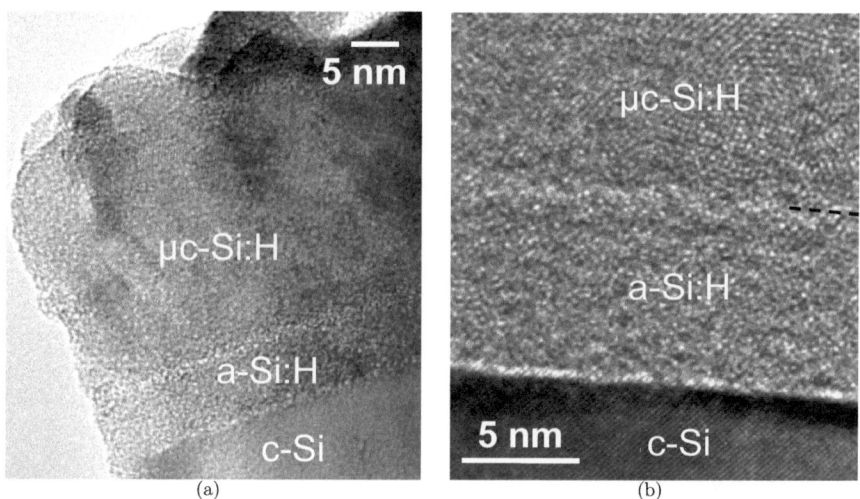

Figure 5.13: *TEM (a) and HR-TEM (b) micrographs of thin-film Si layer stacks on flat c-Si for Si HJ solar cell formation consisting of c-Si / i a-Si:H / doped a-Si:H/µc-Si:H layer (here i/p emitter on n c-Si). Sample prepared by a) the cleaved corner method and b) polishing, see Sec. 2.1.2.3.*

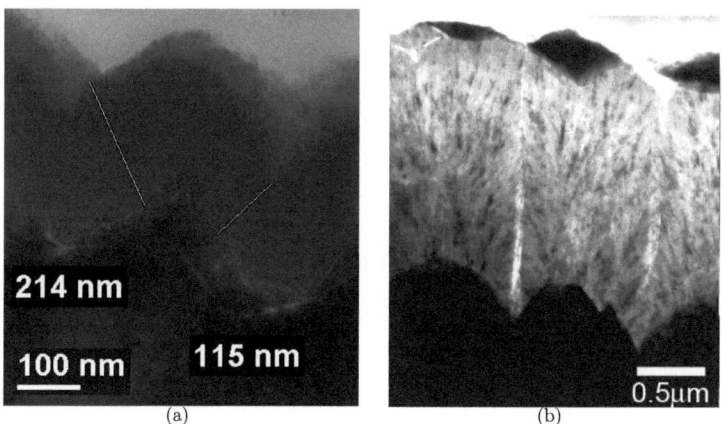

Figure 5.14: *Texture based growth problems of thin-film Si solar cells. a) Non-conformal growth of a-Si:H and b) growth inhomogeneities of µc-Si:H in sharp pinches on LPCVD-ZnO [Pyt09].*

5.5. Textured a-Si:H/c-Si heterojunction solar cells

Figure 5.15 shows the HR-TEM micrograph of the badly performing textured solar cell whose JV-curve is displayed in Fig. 5.12. In contrast to the same layer stack's growth on flat c-Si shown in Fig. 5.13, the intrinsic and the subsequent doped layer are indiscernible and the layer stack's thickness perpendicular to the pyramidal facet (decisive for passivation and thus V_{OC}) is too thin.

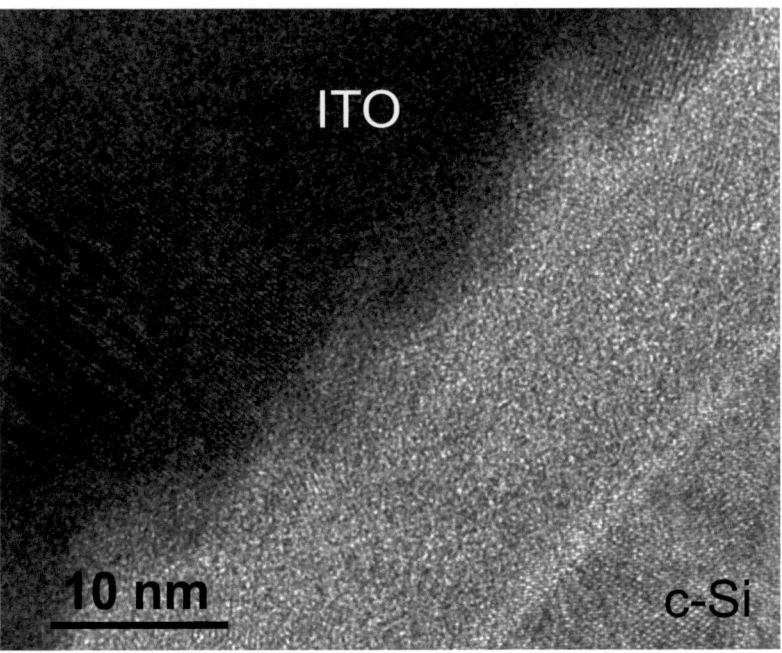

Figure 5.15: *HR-TEM micrograph of ITO / i a-Si:H / doped a-Si:H/µc-Si:H layer stack grown on pyramidal textured c-Si. In contrast to the same layer stack's growth on flat c-Si in Fig. 5.13, the interface between the intrinsic and the doped thin-film Si layer is not visible and the layer stack's thickness perpendicular to the pyramidal facets is reduced.*

We thought the thin-film Si growth to be much less directional, but instead of growing the same thickness as on flat c-Si perpendicular to the pyramidal facets, we grow the same thickness perpendicular to the whole wafer, see geometrical considerations in Fig. 4.34. Figure 4.34 shows that based on the angle between the c-Si's <100> and <111> crystal planes, the growth time would need to be prolonged by a factor of 1.73 to yield

5.5. Textured a-Si:H/c-Si heterojunction solar cells

an equivalent thickness perpendicular to the pyramid facets of the textured c-Si. Although emitter absorption is not enhanced because of the refraction of normally incident light normal to the pyramidal facets, see discussion related to Fig. 4.34, this factor can be reduced to 1.4. The slower growth rate potentially favors the development of the doped layer's particular microstructure on a smaller thickness.

On one single small cell of one single wafer piece (Fig. 2.20(b) for illustration), a V_{OC} of 675 mV with an efficiency of 16.9% ($FF = 69\%$ and $J_{SC} = 36.5$ mA/cm^2) is then effectively reached, but already the second best cell on this wafer piece has a 50 mV lower V_{OC} of only 625 mV. From the discrepancies in the long-wavelength EQE of this best and second best cell in Fig. 5.16 it becomes obvious that back surface passivation is only successful on the best cell, simultaneously increasing the latters V_{OC} as well as its J_{SC}.

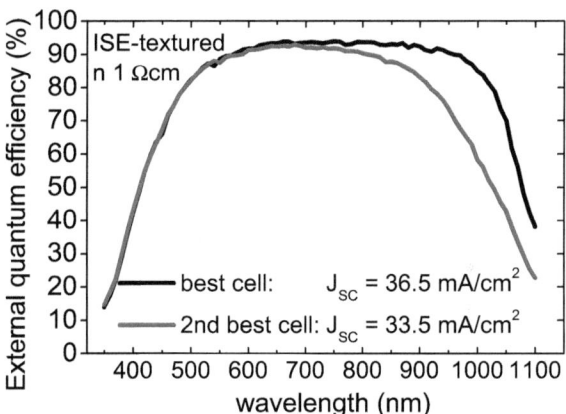

Figure 5.16: *EQE of the best solar cell based on the ISE-textured c-Si wafer compared to the second best cell on the same wafer piece (Fig. 2.20(b)). Only one single small cell has a sufficiently good back surface passivation to yield a high V_{OC}.*

In view of the ultimate goal to make whole wafer Si HJ solar cells, the use of textured wafers yielding homogeneous lifetimes on whole wafers when passivated by a-Si:H is of primordial importance. An overall more regularly textured c-Si wafer (compare Fig. 5.22(a) to Fig. 4.38) can yield pretty homogeneous lifetimes over a large surface after i a-Si:H passivation as shown by the ILM lifetime mapping in Fig. 4.43.

5.5. Textured a-Si:H/c-Si heterojunction solar cells

5.5.2 Influence of the texture morphology

Next, the optimally preconditioned textured c-Si wafers from the HMI [ARK+08] are processed into finished Si HJ solar cells. They are excellently passivated by our standard i a-Si:H layer, implying about the same high V_{OC}s largely over 700 mV as on flat c-Si, on n- as well as on p-type, see Sec. 4.7. Figure 5.17 shows the resulting cells JV- and EQE-characteristics.

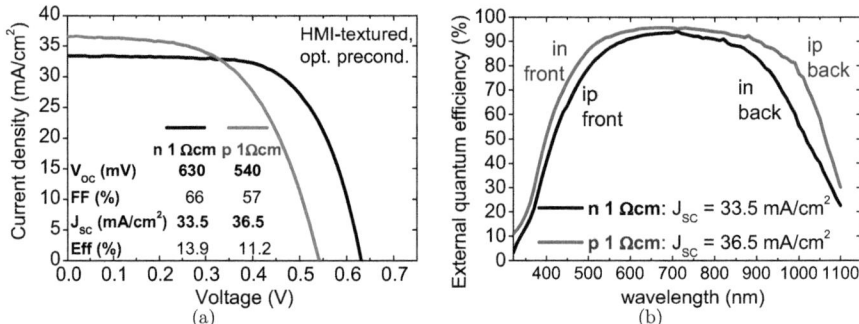

Figure 5.17: *HJ solar cell fabrication on optimally preconditioned textured c-Si from HMI on 1 Ωcm n- and p-type c-Si. a) JV- and b) EQE-characteristics. While i a-Si:H passivation results on these wafers are promising (Sec. 4.7), solar cell results are deteriorated by the BSF layer stack's growth on textured c-Si.*

The high expectations from i a-Si:H passivation are by far not met by the finished Si HJ solar cells. From the 90 mV lower V_{OC} on p-type as compared to n-type c-Si on one hand, but the much higher long-wavelength EQE leading to a J_{SC} discrepancy of 3 mA/cm^2 on the other hand, we conclude that the n-layer growth induces interface defects. The higher short-wavelength EQE of the cell based on p-type c-Si partially results from i) lower absorption in the i/n than in the i/p stack (Fig. 2.1(b)) and ii) probably also from a higher photoelectrical activity due to a lower defect density in n- as compared to p-type a-Si:H/μc-Si:H (generally both supposed to be photoelectrically inactive).

The hypothesis of the n-layer growth inducing interface defects, as stated after Si HJ solar cell fabrication on optimally preconditioned HMI-textured c-Si of both doping types (Fig. 5.17) is confirmed by lifetime measurements on emitter and BSF layer stacks grown symmetrically on the ISE-textured wafer (Sec. 4.7). The increasing i) of the i-layer thickness to

5.5. Textured a-Si:H/c-Si heterojunction solar cells

better protect the i a-Si:H/c-Si interface against the ion-bombardment of the n a-Si:H/µc-Si:H layer's growth, and ii) the increase of the thickness of the n a-Si:H/µc-Si:H layer to permit its specific microstructure's development (that ensures its efficient doping and thus its field-effect passivation (Sec. 2.3.4)), effectively results in higher measured lifetimes (Fig. 4.45). TEM examination of a solar cell containing such a BSF layer stack reveals the requested abrupt crystallographic interfaces between c-Si, a-Si:H and a-Si:H/µc-Si:H on the pyramidal facets, see HR-TEM micrograph in Fig. 5.18(a). Towards the bottom of the pyramidal valley (again Fig. 5.18(a)), the different layers are less clearly discernible. The HR-TEM micrograph in Fig. 5.18(b) confirms the initial epitaxial growth at the bottom of the pyramidal valley, that was already observed for the emitter layer stack in Sec. 4.7.

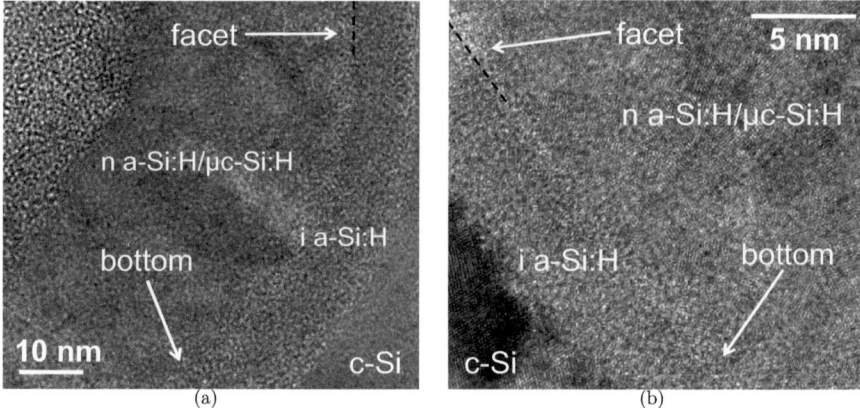

Figure 5.18: *HR-TEM micrographs of a i/n BSF layer stack grown thicker for improved back surface passivation. Transition between pyramidal valley bottom and facet showing a) layer stacks of clearly discernible microstructures on the facets in the upper right. The less clearly discernible layers in the lower left of Fig. a) result as visible in the lower right of Fig. b), from epitaxial growth at the bottom of the pyramidal valleys.*

The FF of such a solar cell suffers from current transport problems through such a thick i a-Si:H passivation layer that is, in addition still locally "short-circuited" because of epitaxial growth at the bottom of the pyramidal valleys. However, the thicker i/n layer stack's passivation is still rather low compared to i a-Si:H passivation (again Fig. 4.45). The

5.5. Textured a-Si:H/c-Si heterojunction solar cells

development of a lower impacting n-layer, i.e. omitting the pre-deposition hydrogen treatment while introducing an intercalated CO_2-treatment to grow a high-conductive layer over a limited thickness, as is done for the p-layer (Sec. 2.3.4), called n_2, finally improves the lifetime of the i/n test structure but decreases its field-effect passivation (again Fig. 4.45). This is because of the n_2-layer's lower conductivity. After successful symmetrical BSF passivation, this new n_2-layer is used in stack with i a-Si:H for solar cell fabrication on the wafer textured with large pyramids by Solarworld whose i a-Si:H passivation induced V_{OC}-limit is reaching a very high value of 720 mV (Sec. 4.7). Compared to the second best solar cell on the ISE-textured wafer (the best is not considered as it was not even reproduced on the same wafer piece), the use of the n_2-layer greatly improves the solar cell V_{OC} and J_{SC}, both by 6%, see JV-curves in Fig. 5.19.

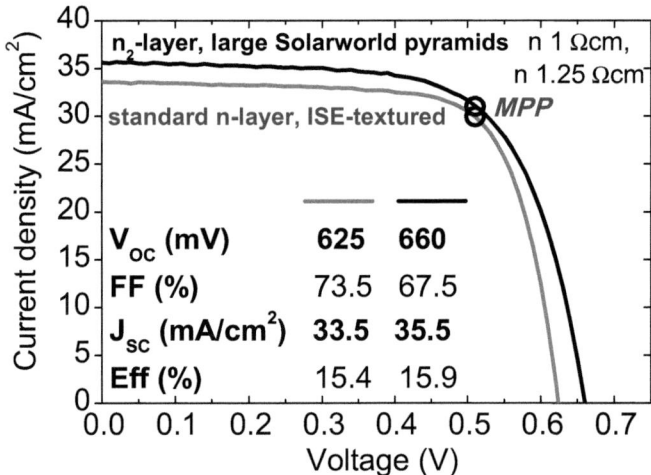

Figure 5.19: JV-curve of a Si HJ solar cell based on the newly developed n_2-layer (less detrimental) and the by Solarworld with regular large pyramids textured wafer compared to the second best solar cell based on ISE-textured c-Si (the very best is disregarded as its reproducibility is even not realized on the same wafer piece). The standard n-layer gives the best results on flat c-Si.

Also the reproducibility on the same wafer piece is much better. However, the absolute gain in efficiency is only 0.5% because of the lowered FF when using the n_2- instead of the n-layer in the BSF layer stack on

5.5. Textured a-Si:H/c-Si heterojunction solar cells

n-type textured c-Si. As shown in Fig. 4.45, the field-effect passivation of the i/n_2 layer stack is lower than the field-effect passivation of the i/n layer stack because of the formers' lower conductivity. Additional current transport problems arise from the thin oxide layer shown to be formed by the CO_2-treatment at the i/n_2 interface [Per01]. Whereas for a-Si:H solar cells, with a lower J_{SC}, the current extraction by tunneling is only hindered for long CO_2-treatment times, more current must be extracted through the Si HJ solar cell's BSF layer stack. In contrast, thanks to the much higher band bending at the emitter side, carrier tunneling is favored towards the standard p-layer that also uses an intercalated CO_2-treatment forming a thin oxide and whose current transport properties (Sec. 5.2) thus should also suffer from the presence of a thin oxide. Possibly in our p-type c-Si based solar cells the current transport properties through the i/p BSF layer stack suffer from the standard use of a CO_2-treatment, that adds up to the less favorable i a-Si:H / p c-Si interface recombination's injection level-dependance yielding the observed lower FFs. In terms of realizing the optimum amount of tunneling for Si HJ solar cell performances (Sec. 5.2), i.e. the highest $V_{OC} \times FF$-product, the combined use of CO_2- (formation of an ultrathin oxide layer) and H_2-treatments (i a-Si:H layer bandgap increase but danger of defect creation by ion bombardment, because the following doped layer grows highest conductive when the i a-Si:H surface is treated by a high power/low pressure H_2-plasma [Per01]) at different locations within the Si HJ solar cell fabrication process should be optimized.

As a consequence, the implementation of the n_2-layer instead of the standard n-layer for solar cell fabrication on flat c-Si brings no benefit, because it decreases FF. However, it should be noted that the ITO_2-layer that is used together with all n_2-layers leads also together with the standard n-layer on flat c-Si to lower FFs than the ITO_1-layer, probably because of its lower conductivity (Sec. 2.4.1).

The HR-TEM micrograph in Fig. 5.20(a) confirms that on the pyramid facets the i a-Si:H and the n_2-layer can be well distinguished. The thickness of the i/n_2 layer stack is reduced as compared to the standard i/n layer stack (Fig. 5.18(a)). Figure 5.20(b) shows that there is still epitaxial growth at the bottom of the pyramidal valleys and that the n_2-layer's surface is very rough.

5.5. Textured a-Si:H/c-Si heterojunction solar cells

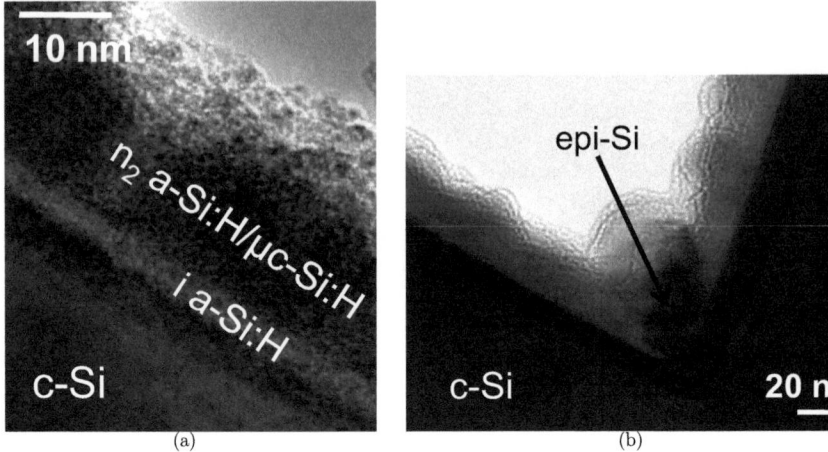

Figure 5.20: *TEM micrographs of a Si HJ solar cell's BSF layer stack containing the newly developed n_2-layer. a) The i a-Si:H and the n_2-layer are clearly discernable, but b) epitaxial growth at the bottom of the pyramidal valleys is still observed and additionally the n_2-layer has a high surface roughness.*

5.5.3 Pyramidal valley rounding

From the passivation decrease observed when an *i* a-Si:H layer's growth conditions leads to local epitaxy in the pyramidal valley (Sec. 4.7), we suggest that this cell's V_{OC} is limited by the locally depassivated areas at the bottom of the pyramidal valleys. In the case of thick micrometer range μc-Si:H layers, the observed growth inhomogeneities in sharp pinches in Fig. 5.14(b) deteriorate the μc-Si:H solar cell's performances [PVSB+08]. These sharp pinches originate from the V-shaped surface morphology of LPCVD-ZnO used as front TCO electrode to doped layers, that can be transformed into a U-shaped surface morphology by means of a plasma treatment. As a consequence, these growth inhomogeneities are no longer observed and solar cells perform much better. In analogy, to avoid epitaxial growth in the c-Si's pyramidal valleys, we intend to round these valleys. For this, we use a CP133 solution consisting of a mixture of HF (50% in H_2O DI), HNO_3 (100% fuming) and CH_3COOH (100%), ratio by volume 1 : 3 : 3, used for bright etching, i.e. wet-chemical polishing of c-Si wafers, thus isotropically etch-polishing c-Si. Figure 5.21 shows the wafer

5.5. Textured a-Si:H/c-Si heterojunction solar cells

textured by Solarworld with very small pyramids, Fig. 5.21(a), with its initially sharp pyramidal valleys including very small pyramids and even untextured surfaces, and Fig. 5.21(b) shows the same wafer after only a few seconds of wet-chemical CP133-etching that effectively rounds the pyramidal valleys eliminating all nanopyramids, homogenizing the pyramid sizes, and diminishing the density of nested pyramids. Further prolonged etching starts to round the pyramid tips and eliminates all nested pyramids as shown in Fig. 5.21(c).

The efficiency of co-deposited µc-Si:H solar cells where the substrate consists of the sputter-ZnO-covered small pyramidally textured c-Si shown in Fig. 5.21, increases from under 2% when using the initially textured c-Si containing sharp pinches (Fig. 5.21(a)) to over 7% when rounding the pyramidal valleys (Figs. 5.21(b) and 5.21(c)) [Pyt09]. This efficiency gain is directly linked to a decrease in defective µc-Si:H growth occurring at sharp pinches and traversing the whole µc-Si:H solar cell (Fig. 5.14(b)) [PVSB+08].

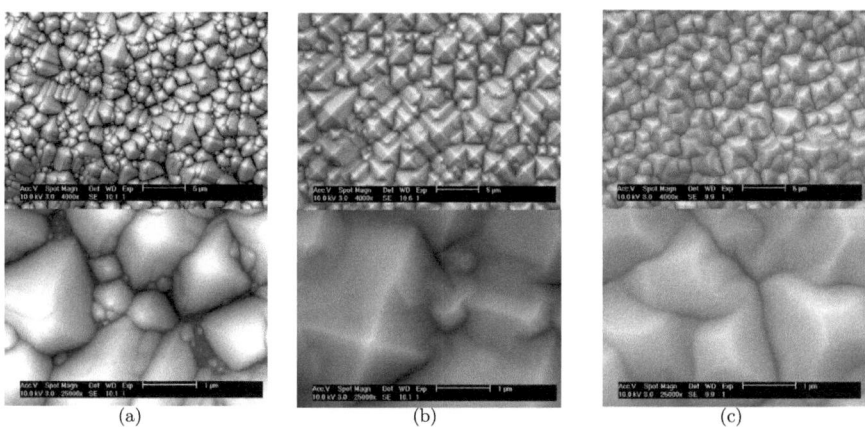

Figure 5.21: *Small pyramidal textured c-Si, a) before and b) and c) after wet-chemical pyramidal valley rounding. Efficiencies of µc-Si:H solar cells grown on these sputter-ZnO-covered c-Si morphologies as substrates, increase from under 2% to over 7% when the pyramidal valleys are rounded (b and c) as compared to the initially textured state (a) also including sharp pinches, nested pyramids and surfaces populated with nanopyramids or not textured at all.*

5.5. Textured a-Si:H/c-Si heterojunction solar cells

Our textured c-Si HJ solar cell's efficiency based on the large pyramids made by Solarworld improves by 7% relative, thanks to rounding of the pyramidal valleys and homogenizing of the textured surface's morphology. This is shown for the case of medium sized pyramids in Fig. 5.22. Note that on the wafers textured by Solarworld with medium to large pyramids, even before pyramidal valley rounding, fewer nested pyramids and much fewer nanopyramids are present than on the small textured and the same pyramid-sized ISE-textured wafers (compare Fig. 5.22(a) to Figs. 5.21(a) and 4.38).

Figure 5.22: *c-Si textured by Solarworld with medium sized pyramids a) before and b) after wet-chemical pyramid valley rounding. The efficiency of Si HJ solar cells is improved by 7% when using the same rounded pyramidal valleys instead of the sharply pinched ones on the wafer with the large Solarworld pyramids.*

The JV-curve of the a-Si:H/c-Si HJ cell based on the large pyramidally textured wafer with rounded pyramid valleys (analogous Fig. 5.22(b) showing medium sized pyramids) is shown in Fig. 5.23. It is here compared to the one of the layers grown with the same process parameters on the same wafer with sharp pyramidal valleys (analogous Fig. 5.22(a) showing medium sized pyramids). Due to this surface modification, solar cells with over 700 mV of V_{OC} are reached. The unchanged low FF confirms the hypothesis of CO_2-treatment induced ultra-thin oxide layer based current transport problems through the i/n_2 BSF layer stack discussed on behalf of Fig. 5.19.

5.5. Textured a-Si:H/c-Si heterojunction solar cells

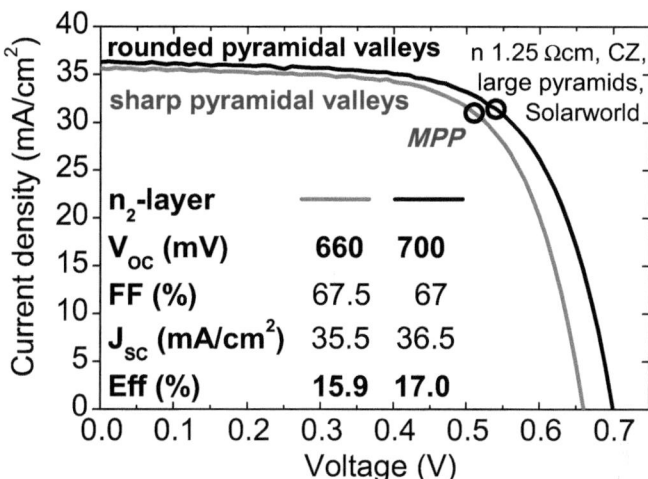

Figure 5.23: *JV-curves of a-Si:H/c-Si HJ solar cells based on n-type CZ c-Si textured by Solarworld with large pyramids and using the texture optimized n_2-layer. Wet-chemical polishing that rounds pyramidal valleys and homogenizes the textured surface's morphology (analogous Fig. 5.22 for medium sized texture) increases V_{OC} from 660 mV to over 700 mV. The additional gain in J_{SC} due to better back surface passivation improves the solar cell efficiency by 7%.*

In addition, current extraction problems could also arise from the rough n_2/ITO interface resulting from the rough n_2-layer's growth (Figs. 5.24(a) and 5.24(b)). The HR-TEM micrographs of the c-Si/i a-Si:H interface in Fig. 5.24(b) and Fig. 5.24(c) reveal also the c-Si's irregular surface. These c-Si surface irregularities resulting from the wet-chemical treatment could possibly be reduced by applying an optimized pyramid rounding treatment. It is known that the isotropicity of the chemical polishing etch critically depends on the exact etch process parameters [Bog67].

5.5. Textured a-Si:H/c-Si heterojunction solar cells

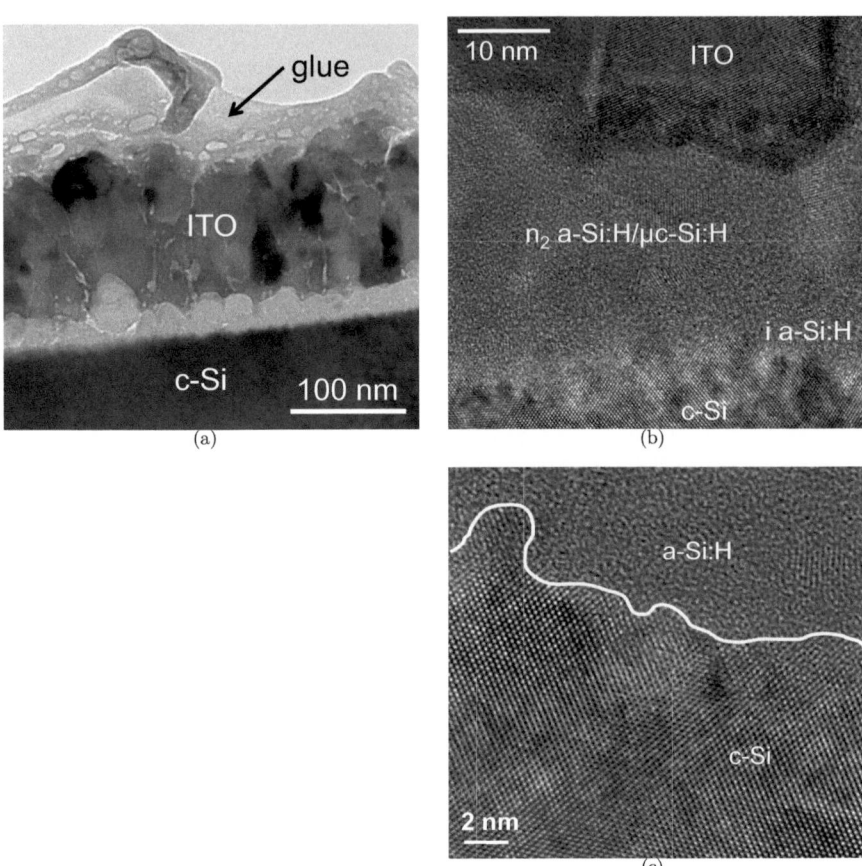

Figure 5.24: *a) and b): TEM and HR-TEM micrograph of the i/n_2 BSF layer stack including the subsequent ITO. Picture taken on a pyramid facet of the best textured Si HJ solar cell of this study having a V_{OC} of over 700 mV. Nonetheless, this cell's efficiency still suffers from a limited FF, which could partially provide from the very rough n_2/ITO interface. b) and c): HR-TEM micrographs reveal additionally the irregular nature of the c-Si's surface resulting from the wet-chemical pretreatment.*

5.5. Textured a-Si:H/c-Si heterojunction solar cells

However, a part of the FF loss is also inherent in the texture based surface increase on same sized cells as compared to flat cells. Figure 5.25 shows that for a flat small cell, doubling the surface leads to a relative FF loss of more than 10% because of series resistance due to the absence of a front metal grid. Because in the case of textured Si HJ solar cell structuring into individual solar cells has to be done by means of a metal mask fixed on top of the c-Si wafer during ITO deposition (as it is no longer possible using a marker (Sec. 2.4.1)), the resulting ill-defined boarder of textured c-Si solar cells decreases FF on textured Si HJ solar cells (as compared to flat c-Si). To the previously mentioned FF measurement problems related to the direct positioning of the measurement probes on ITO, adds that positioning the probes on the textured Si HJ solar cell's leaves back holes in the ITO layer (visible to the eye). Fabrication of larger cells including a front metal grid would simultaneously resolve i) FF measurement uncertainties, ii) measurement probe based ITO perforation problems, and iii) mask related structuring problems.

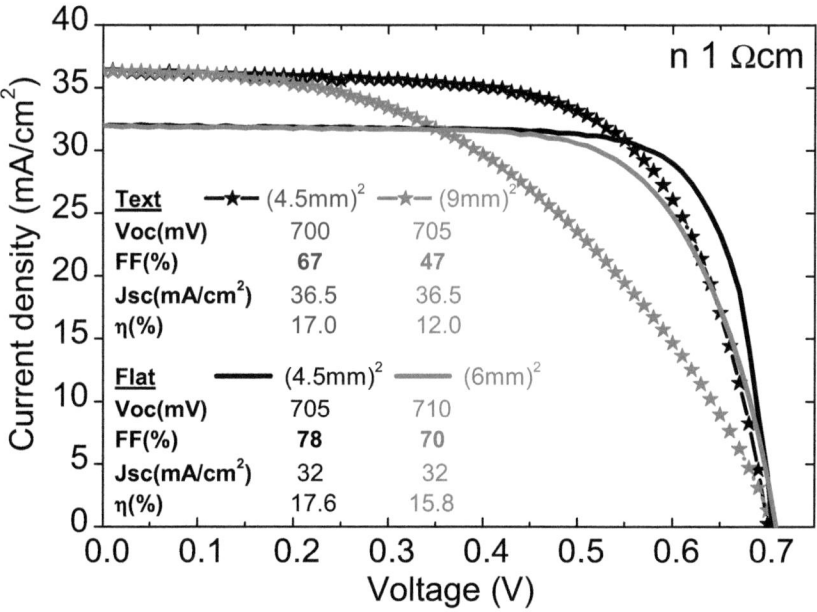

Figure 5.25: *JV-curves of small flat and textured Si HJ solar cells of varying size without front metallization: partially series resistance based FF losses.*

5.5. Textured a-Si:H/c-Si heterojunction solar cells

In terms of J_{SC} there is no improvement, compared to the fortuitous high performing single solar cell based on the ISE-textured c-Si (Fig. 5.16). Their EQE comparison in Fig. 5.26 shows that the high-V_{OC} solar cell based on rounded pyramidal valleys has a lower long-wavelength EQE than the solar cell based on the same sized but more irregularly textured ISE-pyramids (but with a lower V_{OC}). Because both ITO's are not the same (ITO$_1$ for the ISE-textured cell and ITO$_2$ for the high-V_{OC} cell) the EQE* [%] subtracts the contribution of ITO absorption, as measured on flat c-Si (Fig. 2.14) ($EQE^* = \frac{EQE}{(1-A_{ITO})}$).

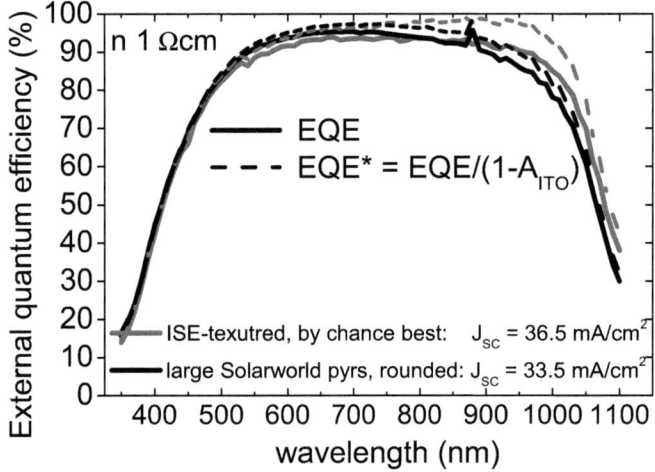

Figure 5.26: *EQE-comparison of a high performance Si HJ solar cell based on the ISE-textured wafer and the high-V_{OC} solar cell based on rounded large pyramidal c-Si textured by Solarworld. Although one can assume from the latter's higher V_{OC} a better back surface passivation, a long-wavelength EQE loss is measured. In EQE*, by subtraction of the ITO absorption, this loss is even more pronounced. In fact, as identified by lifetime measurements (see text), the lower long-wavelength EQE of the solar cell with the higher V_{OC} results from a reduced geometrical light-trapping because of the rounded texture shape in combination with a reduced wafer thickness.*

However, ITO$_2$ is less absorbant than ITO$_1$ and therefore the differences in long-wavelength EQE* are even more pronounced than in EQE. This can be explained as follows: larger pyramids and thinner wafers alone

5.5. Textured a-Si:H/c-Si heterojunction solar cells

do not increase total reflection and transmission as compared to small pyramids on thicker wafers, as confirmed by Fig. 4.40. But for successful geometrical light-trapping, the sharp features of the random pyramids are ideal (Fig. 2.16), while rounded pyramidal valleys lead to enhanced reflection and transmission losses of normally incident light. The latter are the more pronounced, the pyramids are large and the wafer thin. Besides by photospectrometer measurements this can also be verified by lifetime measurements on the solar cell precursor (process path 4 in Fig. 5.5), where the overlapping of the curves acquired in the transient and the generalized QSS mode yields an optical factor of $F = 0.85$ (Sec. 2.1.2.1), which is situated between $F = 1$ measured on bare sharply textured c-Si irrespective of the pyramid's sizes and the wafer thickness and $F = 0.7$ of bare flat c-Si.

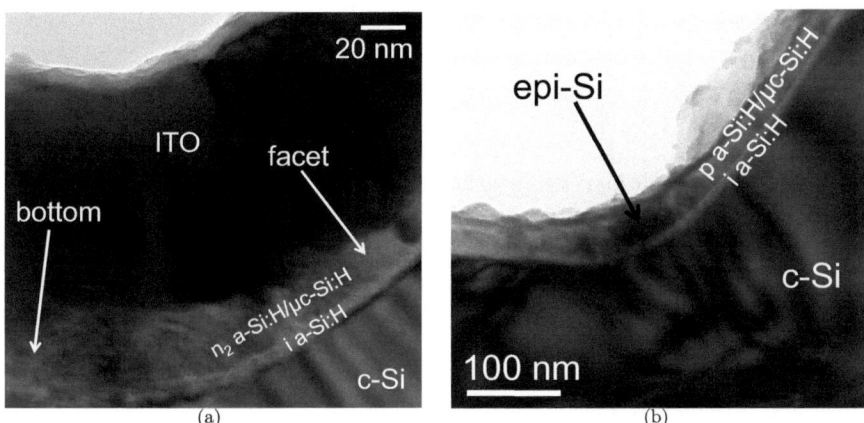

Figure 5.27: *TEM micrographs of a Si HJ solar cell based on large, valley rounded random pyramids fabricated by Solarworld: i/n_2 BSF layer stack including ITO (a) and i/p emitter layer stack (b). In contrast to the stacked layers growth in sharp pyramidal valleys, the i a-Si:H and the doped a-Si:H/μc-Si:H layers are clearly discernible from each other, despite the also in the rounded pyramidal valley present epitaxial growth propagating the Si wafer's crystal orientation through the i a-Si:H into the doped a-Si:H/μc-Si:H layer.*

The TEM micrographs of the BSF and the emitter layer stack in Fig. 5.27(a) respectively 5.27(b), reveal that irrespective of the larger curvature radius, there is still epitaxial growth in the pyramidal valleys. But the i a-Si:H layer and the doped a-Si:H/μc-Si:H layer forming the BSF as well as the emitter layer stack are clearly discernible despite the continuing presence of epitaxy of the doped a-Si:H/µc-Si:H layer in the pyramidal valleys. This is in contrast to the layer stacks grown within the exactly same process parameters on the same wafer but with the initially sharp textured features. Surprisingly, in contrast to flat c-Si, epitaxy can thus occur in the pyramidal valley on a fully amorphous ultra-thin ($\sim$ 2 nm) i a-Si:H layer. As origin, small local areas with direct epitaxy following lateral reorganization might be an explanation.

µc-Si:H growth inhomogeneities coming from sharp ZnO-pinches manifest themselves as "cracks" (Fig. 5.14(b)), and penalize µc-Si:H solar cell performances. The TEM contrast difference forming these cracks are identified as zones of porous material presumed to result from shadowing effects and collision of growth fronts [Pyt09]. By rounding the ZnO pinches, the density of cracks is greatly reduced and corresponding solar cell efficiencies are improved. In c-Si pinches, the growth of an a-Si:H/µc-Si:H transition material layer on top of an ultra-thin a-Si:H layer induces epitaxial growth. SEM micrographs indicate that in sharp c-Si pinches, growth inhomogeneities of thicker a-Si:H passivation layers are also present (Fig. 4.35(a)). By either increasing the a-Si:H layer's hydrogen dilution or by further depositing an a-Si:H/µc-Si:H layer, local epitaxy of this possibly already porous a-Si:H material is observed in sharp c-Si pinches. Possibly stress and/or the wafer's local crystalline orientation play a major role. By increasing the c-Si pinches' curvature radius, epitaxy of the thin-film Si layer stack in these rounded pyramidal valleys is still present on the same amount of surface, but the degree of epitaxy is limited, in such a way that interfaces are clearly discernible.

5.6 Conclusions on amorphous/crystalline silicon heterojunction solar cells

In the case of flat Si HJ solar cells the emitter and BSF layer stack development from glass (thickness, conductivity improvement assisted by crystallinity measurements) to small adaptations in Si heterostructure passivation samples to good Si HJ solar cells is straight-forward. It results in a maximum measured efficiency of 19.1%. HF solution and VHF-PECVD chamber cleanliness issues prove to be more critical for current transport (manifesting itself in low FFs) than they are for solely Si heterostructure passivation. An i a-Si:H layer based (thickness, dangling bond density) trade-off between a high V_{OC} and a high FF is identified.

5.6. Conclusions on amorphous/crystalline silicon heterojunction solar cells

In contrast, on textured c-Si an analogous layer stack development on glass is not valid as texture related growth problems occur. More precisely local epitaxial growth is observed in pyramidal valleys by HR-TEM. The beneficial use of homogenously large textured c-Si results from a reduction of the pyramidal valley density. Although further rounding of these pyramidal valleys does not prevent epitaxial growth within, the degree of epitaxy stays such that the different layers are clearly discernible from each other in TEM micrographs and a solar cell V_{OC} of over 700 mV is reached. Different process parameters are thus found to perform better on textured c-Si because of a primordial importance of the abruptness of all interfaces. This abruptness is much more difficult to reach and no longer observed on textured c-Si with the best flat Si HJ process parameters. Further growth studies are needed to completely avoid epitaxially short-circuited i a-Si:H passivation layers in pyramidal valleys. In this perspective, replacing the a-Si:H/µc-Si:H transition material doped layers by purely amorphous layers has the potential for large device improvements given that a good p a-Si:H/ITO contact can be achieved. Note that TEM studies are never absolutely reliable as only a small zone is observed. This is especially true in the case of the textured c-Si, where sometimes only one single pyramidal valley could be observed.

Despite a texture based J_{SC} gain, the performance of our best textured Si HJ solar cell is still significantly lower than the one of our best Si HJ solar cell on flat c-Si because of not as yet unambiguously identified FF losses, see JV-curve comparison in Fig. 5.28.

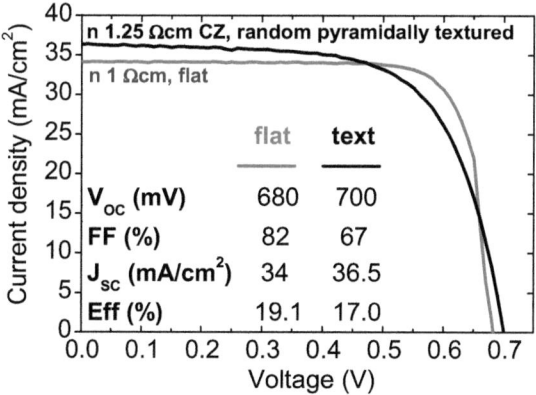

Figure 5.28: JV-curves of the highest-efficient Si HJ solar cells fabricated within this study on flat and on random pyramidally textured c-Si.

Chapter 6
Summary, conclusions and further work

The performance of many silicon devices is limited by electronic recombination losses at the crystalline silicon (c-Si) surface. A proper surface passivation scheme is needed to allow minimizing these losses and thus obtaining high efficiency solar cells. Contrariwise to the standard c-Si surface passivation schemes, amorphous Si (a-Si:H) can not only passivate the c-Si surface but simultaneously form the emitter and back surface field (BSF) of c-Si solar cells.

The standard c-Si surface passivation schemes silicon dioxide (SiO_2) and silicon nitride (SiN_x) are usually modeled by the extended Shockley-Read-Hall (SRH) formalism. It permits the calculation of interface recombination under surface band bending conditions, assuming standard SRH recombination through defects with two states of charge. For the first time, the amphoteric nature of Si dangling bonds is considered in the investigation of the surface passivation properties of a-Si:H on c-Si wafers. For this, we introduce a recombination rate relying on the unique properties of dangling bonds, which possess three states of charge in the otherwise standard surface recombination formalism.

Our model fits well to experimentally measured injection level dependent lifetimes on various combinations of intrinsic (i), microdoped, or internally polarized i a-Si:H layers on all kinds of wafer doping types and levels. It can even fit previously published data for SiO_2 and is thus currently the simplest model that allows an understanding of the largest set of experimentally observed behaviors of passivation layers on c-Si. We are, thus, able to quantify the individual contribution of the two parameters governing passivation, which are, i) the density of interface recombination centers and, ii) the passivation layer's average charge density, governing (by field-effect) the depletion of one carrier type within the wafer. Therefore, we conclude that the growth of i a-Si:H on c-Si leads to a low interface recombination center density and manifests the amphoteric nature of the interface defects. The contribution of field-effect passivation is tuned by varying the average state of charge of the interface's dangling bond recombination centers. a-Si:H passivation of c-Si is more symmetrical, as far as field-effect is concerned, than SiO_2 and SiN_x. According to our model, this is fundamentally related to the more symmetrical microscopic parameters governing the recombination through dangling bonds, which are the neutral capture cross-sections and the above mentioned variable contribution of field-effect passivation. Therefore, the passivation of c-Si with a-Si:H has a broader range of applications compared with SiO_2 and SiN_x, which perform both better on n-type c-Si. Nonetheless, at low injection levels, a-Si:H passivation performs also slightly better on n- than on p-type c-Si. A roadmap emerges for the choice of the optimal wafer type in view of minimal interface recombination and thus maximal Si HJ solar cell performances. Whether this can be related to asymmetric band offsets at the a-Si:H/c-Si interface could be the subject of further investigations. Further work could confirm our interpretation of experimental data with the dangling bond recombination model. By means of surface photovoltage (SPV) measurements, Suwito et al. [SRP+08] found the same amphoteric behavior of the fixed charge density when passivating p- and n-type doped c-Si by intrinsic a-SiC, whereas we arrive at the same conclusion for a-Si:H passivation by modeling the injection level dependency of lifetime data.

The astonishingly low defect density at the i a-Si:H/c-Si interface implies open-circuit voltages (V_{OC}) largely over 700 mV on flat c-Si of all kinds of doping types and levels. Moreover, we observed none of the expected light induced lifetime degradation in our passivation layers (Staebler-Wronski effect). But we found that the passivation of ultra-thin (up to 20 nm thick) a-Si:H layers is not stable against ambient storage due to their outer surface potential modification. In fact, a-Si:H yields even better passivation of <111> oriented c-Si (surface recombination velocity S_{eff} of 1 cm/s), in contrast to SiO_2 and SiN_x which are better passivating <100> oriented c-Si. In view of industrial monocrystalline Si solar cell production, that is based on random pyramidal structures featuring <111> oriented facets, this is an important finding. However, although random pyramids offer <111> oriented facets, the overall passivation of textured c-Si by a-Si:H is very sensitive to growth inhomogeneities in the pyramid bottoms. This imposes an improvement on the texture's quality for a better match with our VHF-PECVD deposition technology.

The excellent a-Si:H/c-Si interface passivation is validated by fabricating complete Si heterojunction (HJ) solar cells. Besides elucidating the a-Si:H/c-Si interface passivation mechanisms, lifetime measurement on i a-Si:H / doped microcrystalline (µc-Si:H) passivation layer stacks allow for a rapid check of emitter and BSF thin-film Si layer stacks suitability for integration in Si HJ solar cells. On flat n-type c-Si, we achieved V_{OC}s up to 710 mV and efficiencies up to 19.1% on (4.5 mm)^2 sized cells. However, the accuracy on fill factor (FF) measurements (and therefore on efficiency) is limited by the absence of a front metallization scheme on our cells (ITO is directly contacted) and by their small sizes. Future standard cells should be at least (1 cm)^2 in size and always feature a front metal grid for current collection. The i a-Si:H / doped µc-Si:H layer stacks optimal for flat c-Si did result in low V_{OC}s when applied directly to textured c-Si wafers. In this situation, the classical thin-film characterization techniques that allowed us to pre-optimize the thin-film Si layer properties for flat Si heterostructures can no longer be fully exploited on textured c-Si wafers. Nonetheless in this situation, transmission electron microscopy (TEM) micrographs permit us to identify device performance limiting factors, such as local epitaxy in pyramidal valleys as the main source of our V_{OC}-loss. By reducing the amount of the resulting local recombination by $i)$ improving the texture quality, $ii)$ adapting the doped a-Si:H/µc-Si:H layers growth conditions and $iii)$ modifying the textured surface's morphology, we succeeded in fabricating Si HJ solar cells with V_{OC}s exceeding 700 mV on n-type textured CZ c-Si. Further studies will be needed to clarify the role of geometry, stress and local crystallographic orientation on local epitaxy. However, when using doped a-Si:H instead of doped a-Si:H/µc-Si:H transition layers, these growth-related problems on textured c-Si could possibly be reduced. Thus, further work could also concentrate on achieving a good p a-Si:H/TCO contact.

Acknowledgments

This work was supported by the National Foundation, by the University of Neuchâtel and by the Axpo Holding AG, Switzerland in the frame of the Axpo Naturstrom Fond.

First of all I thank Prof. Christophe Ballif for having welcomed me into his highly stimulating research group and former Prof. Arvind Shah for having encouraged me to join. Prof. Christophe Ballif advanced not only my scientific success but also my self-confidence during my time at the IMT Neuchâtel.

I owe very much to Dr. Evelyne Vallat-Sauvain for having given me a huge store of knowledge on the base of which I could start my own research. With her passion for scientific research, she encouraged me to not only observe but to try to understand and link these observations. I gratefully thank her for leaving me the freedom to develop my own ideas during the time she guided this work, and for her complete confidence in my efforts.

I greatly appreciated working with Frédéric Freitas, Luc Fesquet, Jérôme Damon-Lacoste and Stefaan De Wolf who one after the other grew the "heterojunction group" out of our two-women team.

Without the various "grey room colleagues" and their valuable technological tips and tricks for thin-film silicon deposition, I would never have been able to achieve appreciable results within a short time. In this respect I highly value also Reto Tscharner's excellent technological maintenance support.

Not least, it was always a pleasure to come to work because of the colleagues with whom I always enjoyed spending time.

For the last part of this work I owe much to the EPFL student Christian Monachon who succeeded in making HR-TEM micrographs of the textured crystalline silicon samples, while helping to guide these examinations' direction in many valuable discussions. I'm also thankful to Dr. Aïcha Hessler-Wyser from the EPFL's CIME for her active participation in this collaboration.

I gratefully acknowledge Adolf Münzer from Solarworld München for preparing wafers with high-quality surface texture and Danick Briand and Laurent Guillot from the IMT's Samlab for cleaning these wafers. Many thanks also to the ISFH and especially Klaus Ramspeck, Jan Schmidt and Karsten Bothe for welcoming me so warmly and spending time measuring with and for me. I very much appreciated the informal scientific exchanges with Lars Korte from the HMI and Matt Page from NREL. I also thank my office colleague Martin Python and the EPFL students Léa Deillon and Guillaume Pasche for providing some of the HR-TEM and SEM micrographs.

The perfect balance of experiment and theory in my thesis I owe largely to Ron Sinton, who invented the lifetime tester used as the main experimental tool in this work. Contrary to many other measurement set-ups, its fast and simple use gives so much information, making it a pleasure to use.

I gratefully acknowledge the excellent complementary inputs of Prof. Christophe Ballif and Dr. Evelyne Vallat-Sauvain for the elaboration of this manuscript. And I also thank my experts Prof. Martin Stutzmann, Prof. Marc Burgelman and Prof. Philipp Aebi for taking the time to read this work and participate in my exam.

In the end, my pleasure at work is always based on personal well-being, and in this respect I sincerely thank Nicolai, Andrea and my family, the latter also for their support of my higher education.

Glossary

A	absorption [%], 6
A_{ITO}	ITO absorption (parasitic) [%], 21
$A_{emitter}$	emitter absorption (parasitic) [%], 22
D	diffusivity [cm^{2}s^{-1}], 10
$D_{it,A}(E)$, $D_{it,D}(E)$	acceptor, donor trap density [cm^{-2}eV^{-1}], 58
D_{it}	surface/interface trap density [cm^{-2}], 52
E_U	correlation energy [eV], 76
E_V, E_C	valence, conduction band edge [eV], 52
E_i	center of the bandgap [eV], 51
E_t	trap energy level [eV], 50
$E_{F,\,p\,a-Si:H}$	p a-Si:H Fermi level [eV], 169
E_{Fn}, E_{Fp}	electron, hole quasi-Fermi level [eV], 75
E_{act}	activation energy [eV], 8
E_{tn}, E_{tp}	demarcation levels, i.e. quasi-Fermi level for electron, hole trap [eV], 74
F	optical constant [ ], 13
FF	fill factor [%], 20
G	generation rate [cm^{-3}s^{-1}], 74
G_L	generation rate [cm^{-3}s^{-1}], 13
H_{dil}	hydrogen dilution [ ], 30
I_L	light intensity [mWcm^{-2}], 7
JV-**curve**	current-voltage characteristic, 5, 18
J_{0x}, J_{01}, J_{02}	recombination current [mAcm^{-2}], 15, 62
J_{SC}	short-circuit current density [mAcm^{-2}], 20
J_{impl}	implied current [mAcm^{-2}], 23
J_{rec}	surface recombination current [mAcm^{-2}], 61
L_{diff}	diffusion length [cm], 10
N	free carrier density (in TCO) [cm^{-3}], 35
$N(E)$	density of states [cm^{-3}eV^{-1}], 76, 80

Glossary

N_D, N_A	donor, acceptor density [cm^{-3}], 52
N_s	surface/interface state density [cm^{-2}], 64, 85
N_t	trap density [cm^{-3}], 51
N_{DB}	bulk dangling bond density [cm^{-3}], 76
Q_G	gate electrode charge density [cm^{-2}], 57
Q_f	fixed insulator charge density [cm^{-2}], 57
Q_s	surface charge density [cm^{-2}], 56
Q_t	localized interface charge density [cm^{-2}], 88
$Q_{DB,bulk}$	total bulk DB charge density [cm^{-3}], 88
Q_{DB}	interface DB charge density [cm^{-2}], 88
Q_{Si}	charge density induced in the c-Si surface [cm^{-2}], 56
Q_{it}	interface trap charge density [cm^{-2}], 57
$Q_{t,bulk}$	localized bulk charge density [cm^{-3}], 88
R	bulk recombination rate [cm^{-3}s^{-1}], 10
R_{Aug}	Auger recombination rate [cm^{-3}s^{-1}], 48
R_{DB}	dangling bond recombination rate [cm^{-3}s^{-1}], 80
R_{SRH}	SRH recombination rate [cm^{-3}s^{-1}], 51
R_{rad}	radiative recombination rate [cm^{-3}s^{-1}], 48
R_{sq}	sheet resistance [$\Omega/\square$], 35
R_{tot}	total recombination rate [cm^{-3}s^{-1}], 158
$Refl$	reflection [%], 6
S	surface recombination velocity [cm/s], 46, 52
SC	silane concentration [%], 30
S_{DB}	dangling bond surface recombination velocity [cm/s], 85
S_{SRH}	SRH surface recombination velocity [cm/s], 52
$S_{eff,2diode}$	effective double-diode surface recombination velocity [cm/s], 62
$S_{eff,DB,c}$	calculated effective dangling bond surface recombination velocity [cm/s], 102
$S_{eff,DB}$	effective dangling bond surface recombination velocity [cm/s], 86
$S_{eff,SRH,c}$	calculated effective SRH surface recombination velocity [cm/s], 65

Glossary

$S_{eff,SRH}$	effective SRH surface recombination velocity [cm/s], 59
$S_{eff,c}$	calculated effective surface recombination velocity [cm/s], 67
$S_{eff,m}$	measured effective surface recombination velocity [cm/s], 64
S_{eff}	effective surface recombination velocity [cm/s], 42, 56
S_{front}, S_{back}	front, back surface recombination velocity [cm/s], 42, 46
S_{gen}	generalized surface recombination velocity [cm/s], 94
T	temperature [K], 8
Tr	transmission [%], 6
U	interface recombination rate [cm^{-2}s^{-1}], 11
U_s	surface/interface recombination rate [cm^{-2}s^{-1}], 46, 52
U_{DB}	dangling bond interface recombination rate [cm^{-2}s^{-1}], 85
U_{SRH}	SRH surface recombination rate [cm^{-2}s^{-1}], 51
V_{OC}	open-circuit voltage [V], 2, 11, 14, 20
W	wafer thickness [cm], 10
ΔE_C, ΔE_V	conduction, valence band offset [eV], 110, 169
$\Delta\sigma$	excess photoconductance [$(\Omega\text{cm})^{-1}$], 12
Δn, Δp	excess electron, hole density [cm^{-3}], 10
Δn_s, Δp_s	excess surface/interface carrier density [cm^{-3}], 46, 52
Δn_{av}, Δp_{av}	average excess electron, hole density [cm^{-3}], 12
α	absorption coefficient [cm^{-1}], 7
ϵ_0	vacuum permittivity [As/Vcm], 58
ϵ_{Si}	relative silicon permittivity [ ], 58
η	sunlight energy conversion efficiency [%], 20
λ	wavelength [nm], 6
μ	free carrier mobility [cm^{2}V^{-1}s^{-1}], 35
μ_n, μ_p	electron, hole mobility in c-Si [cm^{2}V^{-1}s^{-1}], 12

Glossary

ϕ_L	photon flux density [cm^{-2}s^{-1}], 13
ϕ_c	Raman crystallinity factor [%], 10
ϕ_n, ϕ_p	electron, hole quasi-Fermi level [V], 14, 58
ϕ_{ITO}	ITO work function [V], 36, 169
$\phi_{p\,a-Si:H}$	p a-Si:H work function [V], 36, 169
ψ_s	surface potential [V], 58
ρ	resistivity [Ωcm], 35
ρ_A^-, ρ_D^+	ionized acceptor, donor density [cm^{-3}], 87
ρ_{DB}^0, ρ_{DB}^+, ρ_{DB}^-	neutral, positively, negatively charged DB density [cm^{-3}], 80
σ	conductivity [(Ωcm)$^{-1}$], 35
σ_d	dark conductivity (Ωcm)$^{-1}$, 8
σ_n, σ_p	electron, hole capture cross-section [cm^2], 51, 52
σ_n^+, σ_p^-	electron, hole capture cross-section of the charged state [cm^2], 78
σ_n^0, σ_p^0	electron, hole capture cross-section of the neutral state [cm^2], 78
τ	charge carrier recombination lifetime [s], 10, 52
τ_{Aug}	Auger lifetime [s], 46, 48
τ_{bulk}	c-Si bulk lifetime [s], 11, 46, 47
τ_{defect}	defect c-Si bulk lifetime [s], 46, 50
$\tau_{eff,c}$	calculated effective lifetime [s], 67
$\tau_{eff,m}$	measured effective lifetime [s], 64
τ_{eff}	effective lifetime [s], 11
$\tau_{eff}(\Delta n)$-**curve**	injection level dependent lifetime curve, 14
τ_{extr}	extrinsic c-Si bulk lifetime [s], 46, 50
τ_{intr}	intrinsic c-Si bulk lifetime [s], 46, 49
τ_{rad}	radiative carrier lifetime [s], 46, 48
τ_{surf}	surface lifetime [s], 11
µc-Si:H	microcrystalline silicon, 1
µdop	microdoping, 125
$f_A(E)$, $f_D(E)$	electron-, hole-occupation function of acceptor, donor traps [ ], 58
f_{DB}^0, f_{DB}^+, f_{DB}^-	neutral, positively, negatively charged occupation probability [ ], 78
f_{SRH}^0, f_{SRH}^-	neutral, charged SRH occupation probability [ ], 83
i	intrinsic, 2, 30

Glossary

k	Boltzmann constant [J/K], 8
n, p	total electron, hole density [cm^{-3}], 15
n_0, p_0	electron, hole density at thermal equilibrium [cm^{-3}], 14
n_f, p_f	electron, hole free carrier density [cm^{-3}], 74
n_i	intrinsic carrier density [cm^{-3}], 15
n_r	refractive index [ ], 35
n_s, p_s	surface electron, hole density [cm^{-3}], 52
n_t, p_t	localized bandtail electron, hole density [cm^{-3}], 87
n_{0x}, n_{01}, n_{02}	diode ideality factor [ ], 15, 62
q	elementary charge [C], 12
r_p^0, r_n^+, r_n^0, r_p^-	capture rates of neutral, positively and negatively charged DBs [cm^{-3}s^{-1}], 78
v_{th}	thermal velocity [cm/s], 51
a-Si:H	amorphous silicon, 1
BSF	back surface field, 2
c-Si	crystalline silicon, 1
C-V	capacitance-voltage, 46, 64
D^0, D$^+$, D$^-$	neutral, positive, negative DB charge condition, 76
DB	dangling bond, 75
DLTS	deep level transient spectroscopy, 46, 64
e	electron, 77
EQE	external quantum efficiency [%], 21
EQE*	external quantum efficiency without ITO absorption [%], 193
ESR	electron spin resonance, 47
FGA	forming gas anneal, 68
FZ	float zone, 11
h	hole, 77

Glossary

HIT	heterojunction with intrinsic thin-layer, 1, 165
HJ	heterojunction, 1
HR-TEM	high resolution transmission electron microscopy, 17
HWCVD	hot wire chemical vapor deposition, 28
ILM	infrared camera lifetime mapping, 15
implFF	implied FF [%], 157
implV_{OC}	implied open-circuit voltage [V], 15
IQE	internal quantum efficiency [%], 21
IQE*	internal quantum efficiency without parasitic ITO absorption [%], 21
IQE**	internal quantum efficiency without parasitic ITO and emitter absorption [%], 22
ITO	indium tin oxide, 7
MIS	metal-insulator-semiconductor, 64, 68
MPP	maximum power point, 20
MW-PCD	microwave-detected photoconductance decay, 12
PCD	photoconductance decay, 12
PECVD	plasma enhanced chemical vapor deposition, 28
PL	photoluminescence, 12
PV	photovoltaic, 1
QE	quantum efficiency [%], 5, 23
QSSPC	quasi-steady-state photoconductance, 12
SEM	scanning electron microscope, 16
SEM	scanning electron microscopy, 5
Si	silicon, 1
SiN$_x$	silicon nitride, 2, 71
SiO$_2$	silicon dioxide, 2, 45, 68
SRH	Shockley-Read-Hall, 45, 50
SunsV_{OC}	illumination level dependent V_{OC}, 5, 23
TCO	transparent conductive oxide, 34, 35

Glossary

TEM transmission electron microscopy, 2, 17

VHF very high frequency, 28
VHF-PECVD very high frequency plasma enhanced chemical vapor deposition, 29

Bibliography

[Abe99] A.G. Aberle. *Crystalline silicon solar cells: Advanced surface passivation and analysis*. Centre for Photovoltaics Engineering, University of New South Wales Publishing and Printing Services, 1999.

[ACD+06] G. Agostinelli, P. Choulat, H.F.W. Dekkers, E. Vermariën, and G. Beaucarne. Rear surface passivation for industrial solar cells on thin substrates. *Proceedings of the 4th World Conference on Photovoltaic Energy Conversion, Hawaii*, 1:1004–1007, 2006.

[ADV+06] G. Agostinelli, A. Delabie, P. Vitanov, Z. Alexieva, H.F.W Dekkers, S. De Wolf, and G. Beaucarne. Very low surface recombination velocities on p-type silicon wafers passivated with a dielectric with fixed negative charge. *Solar Energy Materials and Solar Cells*, 90(18-19):3438–3443, 2006.

[AGW92] A.G. Aberle, S. Glunz, and W. Warta. Impact of illumination level and oxide parameters on Shockley–Read–Hall recombination at the Si-SiO$_2$ interface. *Journal of Applied Physics*, 71(2):4422–4431, 1992.

[AKR+08] H. Angermann, L. Korte, J. Rappich, E. Conrad, I. Sieber, M. Schmidt, K. Hübener, and J. Hauschild. Optimisation of electronic interface properties of a-Si:H/c-Si hetero-junction solar cells by wet-chemical surface pre-treatment. *Thin Solid Films*, 516:6775–6781, 2008.

[ARK+08] H. Angermann, J. Rappich, L. Korte, I. Sieber, E. Conrad, M. Schmidt, K. Hübener, J. Polte, and J. Hauschild. Wet-chemical passivation of atomically flat and structured silicon substrates for solar cell application. *Applied Surface Science*, 254:3615–3625, 2008.

Bibliography

[ATS59] M.M. Atalla, E. Tannenbaum, and E.J. Scheibner. Stabilization of silicon surfaces by thermally grown oxides. *Bell System Technical Journal*, 38:749–783, 1959.

[AVA+04] G. Agostinelli, P. Vitanov, Z. Alexieva, A. Harizanova, H.F.W. Dekkers, S. De Wolf, and G. Beaucarne. Surface passivation of silicon by means of negative charge dielectrics. *Proceedings of the 19th European Photovoltaic Solar Energy Conference, Paris, France*, 1:132–134, 2004.

[BB48] J. Bardeen and W.H. Brattain. The Transistor, A Semi-Conductor Triode. *Physical Review*, 74(2):230–231, 1948.

[BBB+02] R. Brendel, M. Bail, B. Bodmann, J. Kentsch, and M. Schulz. Analysis of photoexcited charge carrier density profiles in Si wafers by using an infrared camera. *Applied Physics Letters*, 80(3):437–439, 2002.

[Bea78] K.E. Bean. Anisotropic etching of silicon. *IEEE Transactions on Electron Devices*, 25(10):1185–1193, 1978.

[Bec39] A. E. Becquerel. Mémoire sur les effets électriques produits sous l'influence des rayons solaires. *Comptes Rendus de l'Académie des Sciences*, 9:561–567, 1839.

[BHvdS+08] J. Benick, B. Hoex, M.C.M. van de Sanden, W.M.M. Kessels, O. Schultz, and S.W. Glunz. High efficiency n-type Si solar cells on Al_2O_3-passivated boron emitters. *Applied Physics Letters*, 92:253504, 2008.

[BJS+85] D.K. Biegelsen, N.M. Johnson, M. Stutzmann, E.H. Poindexter, and P.J. Caplan. Native defects at the Si/SiO_2 interface - amorphous silicon revisited. *Applications of Surface Science*, 22/23:879–890, 1985.

[BKBS00] M. Bail, J. Kentsch, R. Brendel, and M. Schulz. Lifetime mapping of Si wafers by an infrared camera. *Proceedings of the 28th IEEE Photovoltaic Specialists Conference, New York*, pages 99–103, 2000.

[BOD+07] M. Burrows, R. Opila, K. Demirkan, M. Lu, U. Das, S. Bowden, and R. Birkmire. Evaluation of HF treated amorphous

Bibliography

silicon for photoemission determined electronic levels. *Proceedings of the 22nd European Photovoltaic Solar Energy Conference, Milan, Italy*, pages 1726–1729, 2007.

[Bog67] A. F. Bogenschütz. *Ätzpraxis für Halbleiter*. Hanser, 1967.

[Bre03] R. Brendel. *Thin-Film Crystalline Silicon Solar Cells: Physics and Technology*. Wiley-VCH, 2003.

[Bro53] W.L. Brown. n-Type Surface Conductivity on p-Type Germanium. *Physical Review*, 91(3):518–527, 1953.

[BRR03] J. Brody, A. Rohatgi, and A. Ristow. Review and comparison of equations relating bulk lifetime and surface recombination velocity to effective lifetime measured under flash lamp illumination. *Solar Energy Materials and Solar Cells*, 77(3):293–301, 2003.

[BTY+08] H.M. Branz, C.W. Teplin, D.L. Young, M.R. Page, E. Iwaniczko, L. Roybal, R. Bauer, A.H. Mahan, Y. Xu, P. Stradins, T. Wang, and Q. Wang. Recent advances in hot-wire CVD R&D at NREL: From 18% silicon heterojunction cells to silicon epitaxy at glass-compatible temperatures. *Thin Solid Films*, 516(5):743–746, 2008.

[BWHS96] N. Beck, N. Wyrsch, C. Hof, and A. Shah. Mobility lifetime product-A tool for correlating a-Si:H film properties and solar cell performances. *Journal of Applied Physics*, 79(12):9361–9368, 1996.

[CC06] P.J. Cousins and J.E. Cotter. Minimizing lifetime degradation associated with thermal oxidation of upright randomly textured silicon surfaces. *Solar Energy Materials and Solar Cells*, 90(2):228–240, 2006.

[CFP54] D.M. Chapin, C.S. Fuller, and G.L. Pearson. A New Silicon p-n Junction Photocell for Converting Solar Radiation into Electrical Power. *Journal of Applied Physics*, 25:676–677, 1954.

[Cha96] P. Chabloz. *Les couches épaisses en silicium amorphe. Application comme détecteurs de rayons X*. PhD thesis, Département de Microtechnique, EPFL, Lausanne, Switzerland, 1996.

Bibliography

[CLK+90] S.E. Curry, P.M. Lenahan, D.T. Krick, J. Kanicki, and C.T. Kirk. Evidence for a negative electron-electron correlation energy in the dominant deep trapping center in silicon nitride films. *Applied Physics Letters*, 56(14):1359–1361, 1990.

[Cra84] R.S. Crandall. In *Semiconductors and Semimetals, Vol. 21B*, page 257, New York, USA, 1984.

[CWFS87] H. Curtins, N. Wyrsch, M. Favre, and A.V. Shah. Influence of plasma excitation frequency for a-Si:H thin film deposition. *Plasma Chemistry and Plasma Processing*, 7(3):267–273, 1987.

[CWS87] H. Curtins, N. Wyrsch, and A.V. Shah. High-rate deposition of amorphous hydrogenated silicon: effect of plasma excitation frequency. *Electronics Letters*, 23(5):228–230, 1987.

[Dau04] S. Dauwe. *Low-Temperature Surface Passivation of Crystalline Silicon and its Application to the Rear Side of Solar Cells*. PhD thesis, Fachbereich Physik der Universität Hannover, Germany, 2004.

[DBL+08] U.K. Das, M.Z. Burrows, M. Lu, S. Bowden, and R.W. Birkmire. Surface passivation and heterojunction cells on Si (100) and (111) wafers using dc and rf plasma deposited Si:H thin films. *Applied Physics Letters*, 92:063504, 2008.

[DL07] J. Damon-Lacoste. *Vers une ingénierie de bandes des cellules solaires à hétérojonctions a-Si:H/c-Si. Rôle prépondérant de l'hydrogène*. PhD thesis, LPICM - Laboratoire de Physique des Interfaces et Couches Minces, Ecole Polytechnique, Palaiseau Cedex, France, 2007.

[DMM+03] S. Dauwe, L. Mittelstädt, A. Metz, J. Schmidt, and R. Hezel. Low-temperature rear surface passivation schemes for > 20% efficient silicon solar cells. *Proceedings of the 3rd World Conference on Photovoltaic Energy Conversion, Osaka, Japan*, 2, 2003.

[DMMH02] S. Dauwe, L. Mittelstädt, A. Metz, and R. Hezel. Experimental evidence of parasitic shunting in silicon nitride rear surface passivated solar cells. *Progress in Photovoltaics*, 10(4):271–278, 2002.

Bibliography

[DMP07] L. Deillon, C. Monachon, and G. Pasche. Caractérisation de cellules solaires à hétérojonction. Bachelor's thesis, Ecole Polytechnique Fédérale de Lausanne (EPFL), Lausanne, Switzerland, 2007.

[DN62] S. Deb and B.R. Nag. Measurement of Lifetime of Carriers in Semiconductors through Microwave Reflection. *Journal of Applied Physics*, 33(4):1604–1606, 1962.

[DSH02] S. Dauwe, J. Schmidt, and R. Hezel. Very low surface recombination velocities on p- and n- type silicon wafers passivated with hydrogenated amorphous silicon films. *Proceedings of the 29th IEEE Photovoltaic Specialists Conference, New Orleans, Louisiana*, pages 1246–1249, 2002.

[DWB06] S. De Wolf and G. Beaucarne. Surface passivation properties of boron-doped plasma-enhanced chemical vapor deposited hydrogenated amorphous silicon films on p-type crystalline Si substrates. *Applied Physics Letters*, 88:022104, 2006.

[DWK07] S. De Wolf and M. Kondo. Abruptness of a-Si:H/ c-Si interface revealed by carrier lifetime measurements. *Applied Physics Letters*, 90:042111, 2007.

[DWOB08] S. De Wolf, S. Olibet, and C. Ballif. Stretched-exponential a-Si:H/c-Si interface recombination decay. *Applied Physics Letters*, 93(3):32101, 2008.

[ear] http://www.earth-policy.org/Indicators/Solar/2007.htm.

[ENP03] A. Eray, G. Nobile, and F. Palma. Evaluation of the density of states parameters of a-Si:H by AC photoconductivity measurements and numerical simulation. *Materials Science & Engineering B*, 102(1-3):398–402, 2003.

[epf] http://cime.epfl.ch/.

[ES85] W.D. Eades and R.M. Swanson. Calculation of surface generation and recombination velocities at the Si-SiO$_2$ interface. *Journal of Applied Physics*, 58(11):4267–4276, 1985.

[FG68] D.J. Fitzgerald and A.S. Grove. Surface recombination in semiconductors. *Surface Science*, 9(2):347–369, 1968.

Bibliography

[Fis94]   D. Fischer. *Electric field and photocarrier collection in amorphous silicon p-i-n solar cells: Effects of light-induced degradation and low-level i-layer doping.* PhD thesis, Institut de Microtechnique, Université de Neuchâtel, Neuchâtel, Switzerland, 1994.

[FK05]   H. Fujiwara and M. Kondo. Real-time monitoring and process control in amorphous/crystalline silicon heterojunction solar cells by spectroscopic ellipsometry and infrared spectroscopy. *Applied Physics Letters*, 86:032112, 2005.

[fle]   http://www.flexcell.com/.

[Flu95]   R.S. Flueckiger. *Microcrystalline silicon thin films deposited by VHF plasmas for solar cell applications.* PhD thesis, Institut de Microtechnique, Université de Neuchâtel, Neuchâtel, Switzerland, 1995.

[FOVS+07]   L. Fesquet, S. Olibet, E. Vallat-Sauvain, A. Shah, and C. Ballif. High quality surface passivation and heterojunction fabrication by VHF-PECVD deposition of amorphous silicon on crystalline Si: Theory and experiments. *Proceedings of the 22nd European Photovoltaic Solar Energy Conference, Milan, Italy*, pages 1678–1682, 2007.

[FW07]   H. Fujiwara and J.A. Woollam. *Spectroscopic Ellipsometry: Principles and Applications.* John Wiley & Sons, 2007.

[GBRW99]   S.W. Glunz, D. Biro, S. Rein, and W. Warta. Field-effect passivation of the SiO_2-Si interface. *Journal of Applied Physics*, 86(1):683–691, 1999.

[GF66]   A.S. Grove and D.J. Fitzgerald. Surface effects on p-n junctions: Characteristics of surface space-charge regions under non-equilibrium conditions. *Solid-State Electronics*, 9:783–806, 1966.

[GKS74]   M.A. Green, F.D. King, and J. Shewchun. Minority carrier MIS tunnel diodes and their application to electron- and photo-voltaic energy conversion - I. Theory. *Solid-State Electronics*, 17:551–561, 1974.

[GKV98]   A. Goetzberger, J. Knobloch, and B. Voss. *Silicon Solar Cells: Technology and Systems Applications.* Wiley, 1998.

Bibliography

[GMDK88] R.B.M. Girisch, R.P. Mertens, and R.F. De Keersmaecker. Determination of Si-SiO$_2$ interface recombination parameters using a gate-controlled point-junction diode under illumination. *IEEE Transactions on Electron Devices*, 35(2):203–222, 1988.

[GRB+05] M. Garín, U. Rau, W. Brendle, I. Martín, and R. Alcubilla. Characterization of a-Si:H/c-Si interfaces by effective-lifetime measurements. *Journal of Applied Physics*, 98:093711, 2005.

[GSWW] S.W. Glunz, A.B. Sproul, W. Warta, and W. Wettling. Injection-level-dependent recombination velocities at the Si-SiO$_2$ interface for various dopant concentrations. *Journal of Applied Physics*, 75(3).

[GvdOH+08] J.J.H. Gielis, P.J. van den Oever, B. Hoex, M.C.M. van de Sanden, and W.M.M. Kessels. Real-time study of a-Si:H/c-Si heterointerface formation and epitaxial Si growth by spectroscopic ellipsometry, infrared spectroscopy, and second-harmonic generation. *Physical Review B*, 77(20):205329, 2008.

[Hal52] R.N. Hall. Electron-Hole Recombination in Germanium. *Physical Review*, 87(2):387–387, 1952.

[HDH+92] A.A. Howling, J.L. Dorier, C. Hollenstein, U. Kroll, and F. Finger. Frequency effects in silane plasmas for plasma enhanced chemical vapor deposition. *Journal of Vacuum Science & Technology A: Vacuum, Surfaces, and Films*, 10(4):1080–1085, 1992.

[HG60] J. Hilibrand and R.D. Gold. Determination of the impurity distribution in junction diodes from capacitance-voltage measurements. *RCA Review*, 21:245–252, 1960.

[HHL+06] B. Hoex, S.B.S Heil, E. Langereis, M.C.M. van de Sanden, and W.M.M. Kessels. Ultralow surface recombination of c-Si substrates passivated by plasma-assisted atomic layer deposited Al$_2$O$_3$. *Applied Physics Letters*, 89:042112, 2006.

[HK67] S.M. Hu and D.R. Kerr. Observation of Etching of n-Type Silicon in Aqueous HF Solutions. *Journal of the Electrochemical Society*, 114:414, 1967.

Bibliography

[HKS+08] M. Hofmann, S. Kambor, C. Schmidt, D. Grambole, J. Rentsch, S.W. Glunz, and R. Preu. PECVD-ONO: A New Deposited Firing Stable Rear Surface Passivation Layer System for Crystalline Silicon Solar Cells. *Advances in OptoElectronics*, 2008.

[hmi] http://www.hmi.de/bereiche/SE/SE1/projekte/hetero/index.html.

[HS81] R. Hezel and R. Schörner. Plasma Si nitride-A promising dielectric to achieve high-quality silicon MIS/IL solar cells. *Journal of Applied Physics*, 52(4):3076–3079, 1981.

[HSK+08] M. Hofmann, C. Schmidt, N. Kohn, J. Rentsch, S.W. Glunz, and R. Preu. Stack system of PECVD amorphous silicon and PECVD silicon oxide for silicon solar cell rear side passivation. *Progress in Photovoltaics Research and Applications*, 16(6):509–518, 2008.

[HSM80] R. Hezel, R. Schörner, and T. Meisel. Application of amorphous silicon nitride for MIS/inversion layer solar cells. *Proceedings of the 3rd European Photovoltaic Solar Energy Conference, Cannes, France*, pages 866–870, 1980.

[HSS92] J. Hubin, A.V. Shah, and E. Sauvain. Effects of dangling bonds on the recombination function in amorphous semiconductors. *Philosophical Magazine Letters*, 66(3):115–125, 1992.

[HSSP95] J. Hubin, A.V. Shah, E. Sauvain, and P. Pipoz. Consistency between experimental data for ambipolar diffusion length and for photoconductivity when incorporated into the "standard" defect model for a-Si:H. *Journal of Applied Physics*, 78(10):6050–6059, 1995.

[Ien04] D. Iencinella. *Study of amorphous and nanocrystalline silicon thin films for application as emitter layers in heterojunction solar cells.* PhD thesis, Università degli studi di Bologna, Bologna, Italy, 2004.

[ise] http://www.ise.fraunhofer.de/geschaeftsfelder-und-marktbereiche/silicium-photovoltaik.

Bibliography

[JH85] K. Jäger and R. Hezel. A novel thin silicon solar cell with Al_2O_3 as surface passivation. *Proceedings of the 18th IEEE Photovoltaic Specialists Conference, Las Vegas*, pages 1752–1753, 1985.

[Kah76] D. Kahng. A historical perspective on the development of MOS transistors and related devices. *IEEE Transactions on Electron Devices*, 23(7):655–657, 1976.

[KC] M.J. Kerr and A. Cuevas. General parameterization of Auger recombination in crystalline silicon. *Journal of Applied Physics*, 91(4).

[KC02a] M.J. Kerr and A. Cuevas. Recombination at the interface between silicon and stoichiometric plasma silicon nitride. *Semiconductor Science and Technology*, 17(2):166–172, 2002.

[KC02b] M.J. Kerr and A. Cuevas. Very low bulk and surface recombination in oxidized silicon wafers. *Semiconductor Science and Technology*, 17(1):35–38, 2002.

[KCA+07] L. Korte, E. Conrad, H. Angermann, R. Stangl, and M. Schmidt. Overview on a-Si:H/c-Si heterojunction solar cells - Physics and technology. *Proceedings of the 22nd European Photovoltaic Solar Energy Conference, Milan, Italy*, pages 859–865, 2007.

[Ker76] W. Kern. Chemical etching of dielectrics. *Proceedings of the Symposium on Etching for Pattern Definition. H. Hughes and MJ Rand, Eds. Princeton, New Jersey: The Electrochemical Society*, pages 1–18, 1976.

[Ker90] W. Kern. The Evolution of Silicon Wafer Cleaning Technology. *Journal of the Electrochemical Society*, 137(6):1887–1892, 1990.

[Ker02] M.J. Kerr. *Surface, Emitter and Bulk Recombination in Silicon and Development of Silicon Nitride Passivated Solar Cells*. PhD thesis, Australian National University, Canberra, Australia, 2002.

[KFC+01] S. Klein, F. Finger, R. Carius, H. Wagner, and M. Stutzmann. Intrinsic amorphous and microcrystalline silicon by hot-wire-deposition for thin film solar cell applications. *Thin Solid Films*, 395(1-2):305–309, 2001.

[KFC+03] S. Klein, F. Finger, R. Carius, T. Dylla, B. Rech, M. Grimm, L. Houben, and M. Stutzmann. Intrinsic microcrystalline silicon prepared by hot-wire chemical vapour deposition for thin film solar cells. *Thin Solid Films*, 430(1-2):202–207, 2003.

[KFK08] T. Koida, H. Fujiwara, and M. Kondo. Reduction of Optical Loss in Hydrogenated Amorphous Silicon/Crystalline Silicon Heterojunction Solar Cells by High-Mobility Hydrogen-Doped In_2O_3 Transparent Conductive Oxide. *Applied Physics Express*, 1(4):041501, 2008.

[KLK88] D.T. Krick, P.M. Lenahan, and J. Kanicki. Nature of the dominant deep trap in amorphous silicon nitride. *Physical Review B*, 38(12):8226–8229, 1988.

[KLM+00] J.S. Kim, B. Lägel, E. Moons, N. Johansson, I.D. Baikie, W.R. Salaneck, R.H. Friend, and F. Cacialli. Kelvin probe and ultraviolet photoemission measurements of indium tin oxide work function: A comparison. *Synthetic Metals*, 111:311–314, 2000.

[KS85] D.E. Kane and R.M. Swanson. Measurement of the emitter saturation current by a contactless photoconductivity decay method. *Proceedings of the 18th IEEE Photovoltaic Specialists Conference, Las Vegas*, pages 578–583, 1985.

[KSS89] R.R. King, R.A. Sinton, and R.M. Swanson. Doped surfaces in one sun, point-contact solar cells. *Applied Physics Letters*, 54(15):1460–1462, 1989.

[KTMI96] M. Kondo, Y. Toyoshima, A. Matsuda, and K. Ikuta. Substrate dependence of initial growth of microcrystalline silicon in plasma-enhanced chemical vapor deposition. *Journal of Applied Physics*, 80(10):6061–6063, 1996.

[Lan74] D.V. Lang. Deep-level transient spectroscopy: A new method to characterize traps in semiconductors. *Journal of Applied Physics*, 45(7):3023–3032, 1974.

Bibliography

[LC88]   K.L. Luke and L.J. Cheng. A Chemical/Microwave Technique for the Measurement of Bulk Minority Carrier Lifetime in Silicon Wafers. *Journal of the Electrochemical Society*, 135:957, 1988.

[LMFW01]   P.M. Lenahan, T.D. Mishima, T.N. Fogarty, and R. Wilkins. Atomic-scale processes involved in long-term changes in the density of states distribution at the Si/SiO_2 interface. *Applied Physics Letters*, 79(20):3266–3268, 2001.

[LSK06]   C. Longeaud, J.A. Schmidt, and R.R. Koropecki. Determination of semiconductor band gap state parameters from photoconductivity measurements. II. Experimental results. *Physical Review B*, 73(23):235317, 2006.

[MDR+98]   J.D. Moschner, P. Doshi, D.S. Ruby, T. Lauinger, A.G. Aberle, and A. Rohatgi. Comparison of front and back surface passivation schemes for silicon solar cells. *Proceedings of the 2nd World Conference on Photovoltaic Solar Energy Conversion, Vienna, Austria*, pages 1894–1897, 1998.

[MGAB08]   K.R. McIntosh, J.H. Guo, M.D. Abbott, and R.A. Bardos. Calibration of the WCT-100 photoconductance instrument at low conductance. *Progress in Photovoltaics Research and Applications*, 16(4):279–287, 2008.

[MHS+00]   D.J. Milliron, I.G. Hill, C. Shen, A. Kahn, and J. Schwartz. Surface oxidation activates indium tin oxide for hole injection. *Journal of Applied Physics*, 87(1):572–576, 2000.

[MM04]   R. Meaudre and M. Meaudre. Method for the determination of the capture cross sections of electrons from space-charge-limited conduction in the dark and under illumination in amorphous semiconductors. *Applied Physics Letters*, 85(2):245–247, 2004.

[MNSC91]   A.H. Mahan, B.P. Nelson, S. Salamon, and R.S. Crandall. Deposition of Device Quality, Low H Content a-Si:H by the Hot Wire Technique. *Journal of Non-Crystalline Solids*, 137/138:657–660, 1991.

[NBA99]   H. Nagel, C. Berge, and A.G. Aberle. Generalized analysis of quasi-steady-state and quasi-transient measurements of car-

Bibliography

rier lifetimes in semiconductors. *Journal of Applied Physics*, 86(11):6218–6221, 1999.

[NHK93]   H.C. Neitzert, W. Hirsch, and M. Kunst. Structural changes of a-Si:H films on crystalline silicon substrates during deposition. *Physical Review B*, 47(7):4080–4083, 1993.

[oer]   http://www.oerlikon.com/solar/.

[PB04]   P. Pohl and R. Brendel. Temperature dependent infrared camera lifetime mapping (ILM). *Proceedings of the 19th European Photovoltaic Solar Energy Conference, Paris, France*, pages 46–49, 2004.

[Per01]   P. Pernet. *Développement de cellules solaires en silicium amorphe de type "n-i-p" sur substrats souples*. PhD thesis, Département de Microtechnique, EPFL, Lausanne, Switzerland, 2001.

[PGR+84]   E.H. Poindexter, G.J. Gerardi, M.E. Rueckel, P.J. Caplan, N.M. Johnson, and D.K. Biegelsen. Electronic traps and P_b centers at the Si/SiO_2 interface: Band-gap energy distribution. *Journal of Applied Physics*, 56(10):2844–2849, 1984.

[PHH+00]   P. Pernet, M. Hengsberger, C. Hof, M. Goetz, and A. Shah. Growth of thin <p> uc-Si:H layers for n-i-p solar cells: Effects of the H_2- or CO_2-plasma treatments. *Proceedings of the 16th European Photovoltaic Solar Energy Conference, Glasgow, UK*, pages 498–501, 2000.

[Pla99]   R. Platz. *Amorphous silicon for optimized multi-junction solar cells: Material study and cell design*. PhD thesis, Institut de Microtechnique, Université de Neuchâtel, Neuchâtel, Switzerland, 1999.

[PR06]   M. Petersen and Y. Roizin. Density functional theory study of deep traps in silicon nitride memories. *Applied Physics Letters*, 89:053511, 2006.

[Pra91]   K. Prasad. *Microcrystalline silicon (μc-Si:H) prepared with very high frequency glow discharge (VHF-GD) process*. PhD thesis, Institut de Microtechnique, Université de Neuchâtel, Neuchâtel, Switzerland, 1991.

Bibliography

[PSH+00] J. Perrin, J. Schmitt, C. Hollenstein, A. Howling, and L. Sansonnens. The physics of plasma-enhanced CVD for large area coating: industrial application to flat panel display and solar cells. *Plasma Physics and Controlled Fusion*, 42(12B):353–363, 2000.

[PT79] J.I. Pankove and M.L. Tarng. Amorphous silicon as a passivant for crystalline silicon. *Applied Physics Letters*, 34(2):156–157, 1979.

[PTB08] H. Plagwitz, B. Terheiden, and R. Brendel. Staebler–Wronski-like formation of defects at the amorphous-silicon–crystalline silicon interface during illumination. *Journal of Applied Physics*, 103:094506, 2008.

[PTTB06] H. Plagwitz, Y. Takahashi, B. Terheiden, and R. Brendel. Amorphous Si/SiN double layers: A low-temperature passivation method for diffused phosphorus as well as boron emitters. *Proceedings of the 21st European Photovoltaic Solar Energy Conference, Dresden, Germany*, pages 688–691, 2006.

[PVSB+08] M. Python, E. Vallat-Sauvain, J. Bailat, D. Dominé, L. Fesquet, A. Shah, and C. Ballif. Relation between substrate surface morphology and microcrystalline silicon solar cell performance. *Journal of Non-Crystalline Solids*, 354:2258–2262, 2008.

[Pyt09] M. Python. *Microcrystalline silicon solar cells: growth and defects*. PhD thesis, Institut de Microtechnique, Université de Neuchâtel, Neuchâtel, Switzerland, 2009.

[Rö3] M. Rösch. *Experimente und numerische Modellierung zum Ladungsträgertransport in a-Si:H/c-Si Heterodioden*. PhD thesis, Fakultät für Mathematik und Naturwissenschaften der Carl von Ossietzky Universität Oldenburg, Germany, 2003.

[RD79] R.R. Razouk and B.E. Deal. Dependence of Interface State Density on Silicon Thermal Oxidation Process Variables. *Journal of the Electrochemical Society*, 126(9):1573–1581, 1979.

Bibliography

[RK28]   C.V. Raman and K.S. Krishnan. A new type of secondary radiation. *Nature*, 121(3048):501–502, 1928.

[RNR98]   A. Rohatgi, S. Narasimha, and D.S. Ruby. Effective passivation of the low resistivity silicon surface by a rapid thermal oxide/PECVD silicon nitride stack and its application to passivated rear and bifacial Si solar cells. *Proceedings of the 2nd World Conference on Photovoltaic Solar Energy Conversion, Vienna, Austria*, pages 1566–1569, 1998.

[SA99]   J. Schmidt and A.G. Aberle. Carrier recombination at silicon–silicon nitride interfaces fabricated by plasma-enhanced chemical vapor deposition. *Journal of Applied Physics*, 85(7):3626–3633, 1999.

[sam]   http://www-samlab.unine.ch/home.htm.

[san]   http://www.sanyo.com/solar/.

[Sau92]   E. Sauvain. *Caractérisation du silicium amorphe hydrogéné: Mesures du transport ambipolaire*. PhD thesis, Institut de Microtechnique, Université de Neuchâtel, Neuchâtel, Switzerland, 1992.

[SBLW94] M. Schöfthaler, R. Brendel, G. Langguth, and J.H. Werner. High-quality surface passivation by corona-charged oxides for semiconductor surface characterization. *Proceedings of the 1st World Conference on Photovoltaic Energy Conversion, Hawaii*, pages 1509–1512, 1994.

[SC96]   R.A. Sinton and A. Cuevas. Contactless determination of current–voltage characteristics and minority-carrier lifetimes in semiconductors from quasi-steady-state photoconductance data. *Applied Physics Letters*, 69(17):2510–2512, 1996.

[SDGC+95] M. Sebastiani, L. Di Gaspare, G. Capellini, C. Bittencourt, and F. Evangelisti. Low-Energy Yield Spectroscopy as a Novel Technique for Determining Band Offsets: Application to the c-Si(100)/a-Si:H Heterostructure. *Physical Review Letters*, 75(18):3352–3355, 1995.

[sin]   http://www.sintonconsulting.com/.

Bibliography

[SIOS00] K. Sugiyama, H. Ishii, Y. Ouchi, and K. Seki. Dependence of indium–tin–oxide work function on surface cleaning method as studied by ultraviolet and x-ray photoemission spectroscopies. *Journal of Applied Physics*, 87(1):295–298, 2000.

[SJT85] M. Stutzmann, W.B. Jackson, and C.C. Tsai. Light-induced metastable defects in hydrogenated amorphous silicon: A systematic study. *Physical Review B*, 32(1):23–47, 1985.

[SKL+07] M. Schmidt, L. Korte, A. Laades, R. Stangl, C. Schubert, H. Angermann, E. Conrad, and K. von Maydell. Physical aspects of a-Si:H/c-Si hetero-junction solar cells. *Thin Solid Films*, 515(19):7475–7480, 2007.

[SM06] R.A. Sinton and D. Macdonald. *WCT-120 Photoconductance Lifetime Tester and optional Suns-V_{OC} Stage User Manual*, 2006.

[SMB+08] J. Schmidt, A. Merkle, R. Brendel, B. Hoex, M.C.M. van de Sanden, and W.M.M. Kessels. Surface passivation of high-efficiency silicon solar cells by atomic-layer-deposited Al_2O_3. *Progress in Photovoltaics Research and Applications*, 16(6):461–466, 2008.

[SMG74] H. Schlangenotto, H. Maeder, and W. Gerlach. Temperature dependence of the radiative recombination coefficient in silicon. *Physica Status Solidi (a)*, 21(1):357–367, 1974.

[sol] http://www.solarworld.de/.

[SR52] W. Shockley and W.T. Read. Statistics of the Recombinations of Holes and Electrons. *Physical Review*, 87(5):835–842, 1952.

[SRP+08] D. Suwito, T. Roth, D. Pysch, L. Korte, A. Richter, S. Janz, and S.W. Glunz. Detailed study on the passivation mechanism of a-Si_xC_{1-x} for the solar cell rear side. *Proceedings of the 23rd European Photovoltaic Solar Energy Conference, Valencia, Spain*, pages 1023–1028, 2008.

[SSB+03] W. Sparber, O. Schultz, D. Biro, G. Emanuel, R. Preu, A. Poddey, and D. Borchert. Comparison of texturing methods for monocrystalline silicon solar cells using KOH and

Na_2CO_3. *Proceedings of the 3rd World Conference on Photovoltaic Energy Conversion, Osaka*, 2, 2003.

[SSES96] F.M. Schuurmans, A. Schönecker, J.A. Eikelboom, and W.C. Sinke. Crystal-orientation dependence of surface recombination velocity for silicon nitride passivated silicon wafers. *Proceedings of the 25th IEEE Photovoltaic Specialists Conference, Washington D.C.*, pages 485–488, 1996.

[SSS+97] J. Schmidt, F.M. Schuurmans, W.C. Sinke, S.W. Glunz, and A.G. Aberle. Observation of multiple defect states at silicon–silicon nitride interfaces fabricated by low-frequency plasma-enhanced chemical vapor deposition. *Applied Physics Letters*, 71(2):252–254, 1997.

[SSV+04] A.V. Shah, H. Schade, M. Vanecek, J. Meier, E. Vallat-Sauvain, N. Wyrsch, U. Kroll, C. Droz, and J. Bailat. Thin-film silicon solar cell technology. *Progress in Photovoltaics Research and Applications*, 12(23):113–142, 2004.

[ST71] J.G. Simmons and G.W. Taylor. Nonequilibrium Steady-State Statistics and Associated Effects for Insulators and Semiconductors Containing an Arbitrary Distribution of Traps. *Physical Review B*, 4(2):502–511, 1971.

[Str84] R.A. Street. Disorder effects on deep trapping in amorphous semiconductors. *Philosophical Magazine B*, 49(1):15–20, 1984.

[Str85] R.A. Street. Localized states in doped amorphous silicon. *Journal of Non-Crystalline Solids*, 77/78:1–16, 1985.

[Str91] R.A. Street. *Hydrogenated Amorphous Silicon*. Cambridge University Press, 1991.

[sun] http://www.sunpowercorp.com/.

[SW77] D.L. Staebler and C.R. Wronski. Reversible conductivity changes in discharge-produced amorphous Si. *Applied Physics Letters*, 31(4):292–294, 1977.

[Sze85] S.M. Sze. *Semiconductor devices, physics and technology*. Wiley New York, 1985.

Bibliography

[SZR+05] B. Sopori, Y. Zhang, R. Reedy, K. Jones, Y. Yan, M. Al-Jassim, B. Bathey, and J. Kalejs. A comprehensive model of hydrogen transport into a solar cell during silicon nitride processing for fire-through metallization. *Proceedings of the 31st IEEE Photovoltaic Specialists Conference, Lake Buona Vista*, pages 1039–1042, 2005.

[SZT83] R.A. Street, J. Zesch, and M.J. Thompson. Effects of doping on transport and deep trapping in hydrogenated amorphous silicon. *Applied Physics Letters*, 43(7):672–674, 1983.

[TBH+04] T. Trupke, R.A. Bardos, F. Hudert, P. Würfel, J. Zhao, A. Wang, and M.A. Green. Effective excess carrier lifetimes exceeding 100 milliseconds in float zone silicon determined from photoluminescence. *Proceedings of the 19th European Photovoltaic Solar Energy Conference, Paris, France*, pages 758–761, 2004.

[TDRiCC04] S. Tchakarov, U. Dutta, P. Roca i Cabarrocas, and P. Chatterjee. Modeling of reverse bias dark currents in pin structures using amorphous and polymorphous silicon. *Journal of Non-Crystalline Solids*, 338–340:766–771, 2004.

[TFM+95] P. Torres, R. Flückiger, J. Meier, H. Keppner, U. Kroll, V. Shklover, and A. Shah. Very low temperature epitaxial growth of <p>-type silicon for solar cells. *Proceedings of the 13th European Photovoltaic Solar Energy Conference, Nice, France*, pages 1638–1641, 1995.

[TMT08] M. Taguchi, E. Maruyama, and M. Tanaka. Temperature dependence of amorphous/crystalline silicon heterojunction solar cells. *Japanese Journal of Applied Physics*, 47(2):814–818, 2008.

[top] http://www.topsil.com/.

[TYT+07] Y. Tsunomura, Y. Yoshimine, M. Taguchi, T. Kinoshita, H. Kanno, H. Sakata, E. Maruyama, and M. Tanaka. 22%-efficiency HIT solar cell. *Technical Digest of the International PVSEC-17, Fukuoka, Japan*, pages 387–390, 2007.

[TYT+09] Y. Tsunomura, Y. Yoshimine, M. Taguchi, T. Baba, T. Kinoshita, H. Kanno, H. Sakata, E. Maruyama, and M. Tanaka.

Bibliography

Twenty-two percent efficiency HIT solar cell. *Solar Energy Materials and Solar Cells*, 2009.

[vCRR+98] M.W.M. van Cleef, F.A. Rubinelli, R. Rizzoli, R. Pinghini, R.E.I. Schropp, and W.F. van der Weg. Amorphous silicon carbide/crystalline silicon heterojunction solar cells: A comprehensive study of the photocarrier collection. *Japanese Journal of Applied Physics*, 37(7):3926–3932, 1998.

[vCSR98] M.W.M van Cleef, R.E.I Schropp, and F.A. Rubinelli. Significance of tunneling in p$^+$ amorphous silicon carbide n crystalline silicon heterojunction solar cells. *Applied Physics Letters*, 73(18):2609–2611, 1998.

[VJ86] F. Vaillant and D. Jousse. Recombination at dangling bonds and steady-state photoconductivity in a-Si:H. *Physical Review B*, 34(6):4088–4098, 1986.

[VKST81] M. Vanecek, J. Kocka, J. Stuchlik, and A. Triska. Direct measurement of the gap states and band tail absorption by constant photocurrent method in amorphous silicon. *Solid State Communications*, 39(11):1199–1202, 1981.

[VMF+07] M. Vetter, I. Martín, R. Ferre, M. Garín, and R. Alcubilla. Crystalline silicon surface passivation by amorphous silicon carbide films. *Solar Energy Materials and Solar Cells*, 91(2-3):174–179, 2007.

[VSBM+05] E. Vallat-Sauvain, J. Bailat, J. Meier, X. Niquille, U. Kroll, and A. Shah. Influence of the substrate's surface morphology and chemical nature on the nucleation and growth of microcrystalline silicon. *Thin Solid Films*, 485(1-2):77–81, 2005.

[VSDM+06] E. Vallat-Sauvain, C. Droz, F. Meillaud, J. Bailat, A. Shah, and C. Ballif. Determination of Raman emission cross-section ratio in hydrogenated microcrystalline silicon. *Journal of Non-Crystalline Solids*, 352:1200–1203, 2006.

[VSKM+00] E. Vallat-Sauvain, U. Kroll, J. Meier, A. Shah, and J. Pohl. Evolution of the microstructure in microcrystalline silicon prepared by very high frequency glow-discharge using hydrogen dilution. *Journal of Applied Physics*, 87(6):3137–3142, 2000.

Bibliography

[WAC80] C.R. Wronski, B. Abeles, and G.D. Cody. The influence of carrier generation and collection on short-circuit currents in amorphous silicon solar cells. *Solar Cells*, 2:245–259, 1980.

[WIP+05] T.H. Wang, E. Iwaniczko, M.R. Page, D.H. Levi, Y. Yan, V. Yelundur, H.M. Branz, A. Rohatgi, and Q. Wang. Effective interfaces in silicon heterojunction solar cells. *Proceedings of the 31st IEEE Photovoltaic Specialists Conference, Lake Buona Vista*, pages 955–958, 2005.

[WK88] W. Wanlu and L. Kejun. Studies of some properties of mechanical stress in a-Si:H, a-SiN$_x$:H and a-Si:H/a-SiN$_x$:H heterojunction films. *Thin Solid Films*, 165:173–179, 1988.

[wor] http://www.worldofphotovoltaics.com.

[WTS+91] K. Wakisaka, M. Taguchi, T. Sawada, M. Tanaka, T. Matsuyama, T. Matsuoka, S. Tsuda, S. Nakano, Y. Kishi, and Y. Kuwano. More than 16% solar cells with a new "HIT" (doped a-Si / nondoped a-Si / crystalline Si) structure. *Proceedings of the 22nd IEEE Photovoltaic Specialists Conference*, pages 887–892, 1991.

[YAC+86] E. Yablonovitch, D.L. Allara, C.C. Chang, T. Gmitter, and T.B. Bright. Unusually Low Surface-Recombination Velocity on Silicon and Germanium Surfaces. *Physical Review Letters*, 57(2):249–252, 1986.

[YG86] E. Yablonovitch and T. Gmitter. Auger recombination in silicon at low carrier densities. *Applied Physics Letters*, 49(10):587–589, 1986.

[YPW+06] Y. Yan, M. Page, TH Wang, M.M Al-Jassim, H.M. Branz, and Q. Wang. Atomic structure and electronic properties of c-Si/ a-Si:H heterointerfaces. *Applied Physics Letters*, 88:121925, 2006.

[YSEW86] E. Yablonovitch, R.M. Swanson, W.D. Eades, and B.R. Weinberger. Electron-hole recombination at the Si-SiO$_2$ interface. *Applied Physics Letters*, 48(3):245–247, 1986.

[ZWGF98] J. Zhao, A. Wang, M.A. Green, and F. Ferrazza. 19.8% efficient "honeycomb" textured multicrystalline and 24.4%

monocrystalline silicon solar cells. *Applied Physics Letters*, 73(14):1991–1993, 1998.

List of publications

OliDresd06   S. Olibet, E. Vallat-Sauvain, C. Ballif, "Effect of light induced degradation on passivating properties of a-Si:H layers deposited on crystalline Si," *Proceedings of the 21st EU PV Conference*, Dresden, Germany, pp. 1366–1370, 2006.

OliPRB07   S. Olibet, E. Vallat-Sauvain, C. Ballif, "Model for a-Si:H/c-Si interface recombination based on the amphoteric nature of silicon dangling bonds," *Physical Review B*, vol. 76, 035326, 2007.

OliNum07   S. Olibet, E. Vallat-Sauvain, C. Ballif, L. Fesquet, "Recombination through amphoteric states at the amorphous/crystalline silicon interface: modelling and experiment," *Proceedings of NUMOS-workshop (Numerical modelling of thin film solar cells)*, Gent, Belgium, pp. 141–153, 2007.

OliNREL07   S. Olibet, E. Vallat-Sauvain, C. Ballif, L. Korte, L. Fesquet, "Silicon solar cell passivation using heterostructures," *Proceedings of the 17th Workshop on Crystalline Silicon Solar Cells and Modules: Materials and Processes*, Vail, Colorado USA, pp. 130–137, 2007.

FesMil07   L. Fesquet, S. Olibet, E. Vallat-Sauvain, A. Shah, C. Ballif, "High quality surface passivation and heterojunction fabrication by VHF-PECVD deposition of amorphous silicon on crystalline Si: Theory and experiments," *Proceedings of the 22nd EU PV Conference*, Milan, Italy, pp. 1678–1682, 2007.

OliFuk07   S. Olibet, E. Vallat-Sauvain, C. Ballif, L. Korte, L. Fesquet, "Heterojunction solar cell efficiency improvement on various c-Si substrates by interface recombination modelling," *Technical Digest of the 17th International Photovoltaic Science and Engineering Conference*, Fukuoka, Japan, pp. 99–100, 2007.

WoAPL08   S. De Wolf, S. Olibet, C. Ballif, "Stretched-exponential a-Si:H/c-Si interface recombination decay," *Applied Physics Letters*, vol. 93,

032101, 2008.

OliVal08 S. Olibet, C. Monachon, A. Hessler-Wyser, E. Vallat-Sauvain, S. De Wolf, L. Fesquet, J. Damon-Lacoste, C. Ballif, "Textured silicon heterojunction solar cells with over 700 mV open-circuit voltage studied by transmission electron microscopy," *Proceedings of the 23rd EU PV Conference*, Valencia, Spain, pp. 1140–1144, 2008.

Appendix A
Numerical surface potential calculation

A.1 Surface potential calculation: numerical approximation of Ψ_s from Q_{it}, Q_f and Q_G

Listing A.1: *PsisApproxFromQitQfQg.m:* numerical MATLAB-*approximation of the surface potential Ψ_s from Q_{it}, Q_f and Q_G, Eqs. 3.17, 3.19, 3.21 and 3.23*

```
100  data = n2_8i; wafertype = 'n2.8';

     recmodel = 'DB'; Eti = 0; %%Eti only for SRH
     NS = 0.8e9; Qf = -2.2e10;
     rappsign0p0 = 1/20; rappneg0 = 500; rapnpos0 = 500;
105
     switch (wafertype)
         case 'n2.8'
             p0 = 1e20/1.675e15; w = 100*278*1e-6;
             tau_def = 0.0137; OptTrans = 0.7;
110      case 'p2.5'
             p0 = 5.709e15; w = 100*300*1e-6;
             tau_def = 0.037; OptTrans = 0.7; %tau_def = 0.0034;
     end

115  switch (recmodel)
         case 'DB'
             DB = 1;
         case 'SRH'
             DB = 0;
120  end

     Dn_M = data(:,1).';
     tau_effM = data(:,2);

125
     n0 = 1e20/p0; ni = 1e10;

     vthn = 2.3e7; vthp = 1.88e7; %vth = 1e7; vthn=vth;vthp=vth;
```

A.1. Surface potential calculation: numerical approximation of Ψ_s from Q_{it}, Q_f and Q_G

```
      Esi = 11.9*8.85*(1e-12)/100; Eg = 1.120;
130   q = 1.6e-19; k = 1.38e-23; T = 300; B = 1/0.026; %%B = q/kT;

      Dn_C = logspace(12,17);
      nb = n0+Dn_C; pb = p0+Dn_C;
      n1=ni*exp(B*Eti)*ones(1,length(Dn_C));
135   p1=ni*exp(-B*Eti)*ones(1,length(Dn_C));

      sigp0 = 1e-16; sign0 = rappsign0p0*sigp0;
      sigpneg = sigp0 * rappneg0; signpos = sign0 * rapnpos0;

140   Sp0 = vthp * sigp0 * NS; Sn0 = vthn * sign0 * NS;
      %Tp = 1./(vthp * sigp0 * NS.^(3/2)); Tn = 1./(vthn * sign0 * NS.^(3/2));

      phin = -1/B*log(nb./ni); phip = 1/B*log(pb./ni);

145   Psis_start = sign(Qf)*0.001*ones(1,length(Dn_C));

      delta(1)=1e20;

      for a=1:length(Dn_C)
150       for i=2:100
              Psis(a) = Psis_start(a);

              QSi(a) = -sign(Psis(a))*((2*k*T*ni*Esi/(q*q))*...
                  (exp(B*(phip(a)-Psis(a)))-exp(B*phip(a)))...
155               +exp(B*(Psis(a)-phin(a)))-exp(-B*phin(a))...
                  +q*Psis(a)*(p0-n0)/(k*T*ni)))^(1/2);

              ns(a) = nb(a).*exp(B*Psis(a)); ps(a) = pb(a).*exp(-B*Psis(a));

160           if (DB)
                  fpos = (sigpneg*sigp0*ps(a).^2)./(sigpneg*sigp0*ps(a).^2 ...
                      +signpos*sigpneg*ns(a).*ps(a)+sign0*signpos*ns(a).^2);
                  fneg = (signpos*sign0*ns(a).^2)./(sigpneg*sigp0*ps(a).^2 ...
                      +signpos*sigpneg*ns(a).*ps(a)+sign0*signpos*ns(a).^2);
165               Qit(a) = fpos*NS-fneg*NS;
              else
                  fd = (sign0*n1(a)+sigp0*ps(a))/(sign0*(ns(a)+n1(a))...
                      +sigp0*(ps(a)+p1(a)));
                  fa = (sign0*ns(a)+sigp0*p1(a))/(sign0*(ns(a)+n1(a))...
170                   +sigp0*(ps(a)+p1(a)));
                  Qit(a) = fd*NS-fa*NS;
              end

              Qg = 0;
175
              Qtot(a) = QSi(a)+Qit(a)+Qf+Qg;

              delta(i) = abs(Qtot(a));
              if(delta(i)>delta(i-1))
180           else
                  Psis_start(a) = Psis(a) + 0.001*sign(Qtot(a));
              end
          end
      end
185
      if (DB)
          UDB_C = (ns.*Sn0.'+ps.*Sp0.')./((ps./ns).*(sigp0/signpos).*...
              (vthp/vthn)+1+(ns./ps).*(sign0/sigpneg).*(vthn/vthp));
      else
190       UDB_C = (ns.*ps-ni^2)./((ns+n1)./Sp0.'+(ps+p1)./Sn0.');
      end
      Seff_C = (UDB_C./Dn_C).';

195   %%tau_aug and tau_rad Kerr
      R_augC = nb.*pb.*(1.8e-24*n0^(0.65)+6e-25*p0^(0.65)+3e-27*Dn_C.^(0.8));
      tau_augC = Dn_C./R_augC;

      R_radC = 9.5e-15*nb.*pb; tau_radC = Dn_C./R_radC;
200
```

A.2. Surface potential calculation: numerical solution of non-linear equation relating Ψ_s and Q_s

```
        %%1/tau_bulk = 1/tau_def + 1/tau_aug + 1/tau_rad
        tau_bulkC = 1./(1./tau_def+1./tau_augC+1./tau_radC);

        tau_effC = 1./(1./tau_bulkC+2*Seff_C.'/w);
205
        figure;
        loglog(Dn_C,tau_bulkC); hold on;
        loglog(Dn_C,tau_augC,'k+'); hold on;
210     loglog(Dn_C,tau_radC,'r+'); hold on;
        loglog(Dn_C,tau_def*ones(length(tau_augC),1),'g+'); hold on;
        loglog(Dn_C,tau_effC,'b-','LineWidth',2); hold on;
        loglog(Dn_M,tau_effM,'b*'); hold on;
        axis([min(Dn_C), max(Dn_C), 1e-5, 1e-2])
215     xlabel('\Deltan [cm^{-3}]')
        ylabel('\tau [s]')
        title(['\tau_{eff}: \tau_{defect}=',sprintf('%0.2g',tau_def),...
            's, N_S=',sprintf('%0.2g',NS),'cm^{-2}, Q_f=',...
            sprintf('%0.2g',Qf),'cm^{-2}'])
220     legend('\tau_{bulk Kerr}','\tau_{Auger}','\tau_{radiative}',...
            '\tau_{defect}','\tau_{eff,calc}','\tau_{eff,mes}',4)
```

A.2 Surface potential calculation: numerical solution of non-linear equation relating Ψ_s and Q_s

Listing A.2: *PsisNumSolvFromQs.m: numerical solution of non-linear Eq. 3.24 relating the surface potential Ψ_s and the surface charge Q_s by* MATLAB

```
100     data = p2_5i; wafertype = 'p2.5';

        recmodel = 'DB'; Eti = 0; %%Eti only for SRH
        NS = 1.4e9; Qf = 1.8e10;
        rappsign0p0 = 1/20; rappneg0 = 500; rappos0 = 500;
105
        switch (wafertype)
            case 'n2.8'
                p0 = 1e20/1.675e15; w = 100*278*1e-6;
                tau_def = 0.0137; OptTrans = 0.7;
110         case 'p2.5'
                p0 = 5.709e15; w = 100*300*1e-6;
                tau_def = 0.037; OptTrans = 0.7; %tau_def = 0.0034;
        end

115     switch (recmodel)
            case 'DB'
                DB = 1;
            case 'SRH'
                DB = 0;
120     end

        Dn_M = data(:,1).';
        tau_effM = data(:,2);

125
        global C1 C2 C3 C4;
```

A.2. Surface potential calculation: numerical solution of non-linear equation relating Ψ_s and Q_s

```
          n0 = 1e20/p0; ni = 1e10;

130       vthn = 2.3e7; vthp = 1.88e7; %vth = 1e7; vthn=vth;vthp=vth;
          Esi = 11.9*8.85*(1e-12)/100; Eg = 1.120;
          q = 1.6e-19; k = 1.38e-23; T = 300; B = 1/0.026; %%B = q/kT;

          Dn_C = logspace(12,17);
135       nb = n0+Dn_C; pb = p0+Dn_C;
          n1=ni*exp(B*Eti)*ones(1,length(Dn_C));
          p1=ni*exp(-B*Eti)*ones(1,length(Dn_C));

          sigp0 = 1e-16; sign0 = rappsign0p0*sigp0;
140       sigpneg = sigp0 * rappneg0; signpos = sign0 * rapnpos0;

          Sp0 = vthp * sigp0 * NS; Sn0 = vthn * sign0 * NS;
          %Tp = 1./(vthp * sigp0 * NS.^(3/2)); Tn = 1./(vthn * sign0 * NS.^(3/2));

145       phin = -1/B*log(nb./ni); phip = 1/B*log(pb./ni);

          QS = Qf;

          %%find Psis:
150       %%function F = PsisCalc(x)
          %%F = C1 * exp(-x) + C2 * exp(x) + C3 * x + C4
          C11 = exp(B*phip); C22 = exp(-B*phin); C3 = (p0-n0)/ni;
          C44 = -q*q/(2*k*T*ni*Esi)*QS.*QS-C11-C22;

155       %%define startpoint
          %%Psis positive for positive surface charge
          x0 = sign(QS)*5;

          for a=1:length(Dn_C)
160           C1 = C11(a); C2 = C22(a); C4 = C44(a);

              %%call function "Psiscalc"
              [x, resnorm] = lsqnonlin(@PsisCalc, x0);

165           %%save x in vector
              Psis(a) = x./B;
          end

          ns = nb.*exp(B*Psis); ps = pb.*exp(-B*Psis);
170
          if (DB)
              UDB_C = (ns.*Sn0.'+ps.*Sp0.')./((ps./ns).*(sigp0/signpos).*...
                  (vthp/vthn)+1+(ns./ps).*(sign0/sigpneg).*(vthn/vthp));
          else
175           UDB_C = (ns.*ps-ni^2)./((ns+n1)./Sp0.'+(ps+p1)./Sn0.');
          end
          Seff_C = (UDB_C./Dn_C).';

180       %%tau_aug and tau_rad Kerr
          R_augC = nb.*pb.*(1.8e-24*n0^(0.65)+6e-25*p0^(0.65)+3e-27*Dn_C.^(0.8));
          tau_augC = Dn_C./R_augC;

          R_radC = 9.5e-15*nb.*pb; tau_radC = Dn_C./R_radC;
185
          %%1/tau_bulk = 1/tau_def + 1/tau_aug + 1/tau_rad
          tau_bulkC = 1./(1/tau_def+1./tau_augC+1./tau_radC);

          tau_effC = 1./(1./tau_bulkC+2*Seff_C.'/w);
190

          figure;
          loglog(Dn_C,tau_bulkC);hold on;
          loglog(Dn_C,tau_augC,'k+');hold on;
195       loglog(Dn_C,tau_radC,'r+');hold on;
          loglog(Dn_C,tau_def*ones(length(tau_augC),1),'g+');hold on;
          loglog(Dn_C,tau_effC,'b-','LineWidth',2); hold on;
          loglog(Dn_M,tau_effM,'b*'); hold on;
```

A.2. Surface potential calculation: numerical solution of non-linear equation relating Ψ_s and Q_s

```
      axis ([min(Dn_C), max(Dn_C), 1e-5, 1e-2])
200   xlabel('\Deltan [cm^{-3}]')
      ylabel('\tau [s]')
      title ([ '\tau_{eff}: \tau_{defect}=',sprintf('%0.2g',tau_def) ,...
          's, N_S=',sprintf('%0.2g',NS),'cm^{-2}, Q_f=' ,...
          sprintf('%0.2g',Qf),'cm^{-2}'])
205   legend('\tau_{bulk Kerr}','\tau_{Auger}','\tau_{radiative}' ,...
          '\tau_{defect}','\tau_{eff,calc}','\tau_{eff,mes}',4)
```

Listing A.3: *PsisCalc.m*

```
100   %% C1...C4 must be defined as global variables!
      function F = PsisCalc(x)
      global C1 C2 C3 C4;
      F = C1 * exp(-x) + C2 * exp(x) + C3 * x + C4;
```

I want morebooks!

Buy your books fast and straightforward online - at one of world's fastest growing online book stores! Environmentally sound due to Print-on-Demand technologies.

Buy your books online at
www.morebooks.shop

Kaufen Sie Ihre Bücher schnell und unkompliziert online – auf einer der am schnellsten wachsenden Buchhandelsplattformen weltweit! Dank Print-On-Demand umwelt- und ressourcenschonend produziert.

Bücher schneller online kaufen
www.morebooks.shop

info@omniscriptum.com
www.omniscriptum.com

Printed by Books on Demand GmbH, Norderstedt / Germany